AGAINST
THE
SEAS

AGAINST THE SEAS

SAVING CIVILIZATIONS FROM RISING WATERS

MARY SODERSTROM

DUNDURN
PRESS

Publisher and acquiring editor: Kwame Scott Fraser | Editor: Dominic Farrell
Cover and interior designer: Karen Alexiou
Cover image: water: shutterstock.com/natashanast; city: rawpixel.com/Freepik

Library and Archives Canada Cataloguing in Publication

Title: Against the seas : saving civilizations from rising waters / Mary Soderstrom.
Names: Soderstrom, Mary, 1942- author.
Description: Includes bibliographical references and index.
Identifiers: Canadiana (print) 2022043252X | Canadiana (ebook) 20220432538 | ISBN 9781459750487 (softcover) | ISBN 9781459750494 (PDF) | ISBN 9781459750500 (EPUB)
Subjects: LCSH: Sea level—Environmental aspects. | LCSH: Sea level—Environmental aspects—History. | LCSH: Sea level—Forecasting. | LCSH: Flood forecasting. | LCSH: Flood control. | LCSH: Flood damage prevention. | LCSH: Climatic changes.
Classification: LCC GC89 .S63 2023 | DDC 551.45/8—dc23

We acknowledge the support of the Canada Council for the Arts and the Ontario Arts Council for our publishing program. We also acknowledge the financial support of the Government of Ontario, through the Ontario Book Publishing Tax Credit and Ontario Creates, and the Government of Canada.

Dundurn Press
1382 Queen Street East
Toronto, Ontario, Canada M4L 1C9
dundurn.com, @dundurnpress 𝕏 f ⊙

*This one is particularly for Jeanne Nivon and Thomas and Louis Soderstrom,
but also for all children who will grow up in a world with rising seas.*

Our Immortality

From ages long, long past
Great thundering masses
Of green-blue water have crashed
Against this shore,
Each retreating wave carrying with it
Part of the eternal rock
To lands far off —
To lands unborn.

So has it been
Since earth began;
So will it be
When little man has gone his little way.
The coast will change;
New harbours shelter other ships,
New cliffs protect new, wiser nations
From ravages of storm and surf.

And you and I like all mankind, shall fade,
Not see the golden dawn of better days.
And yet our love, like life, remains —
Seen in the calm green depths, the cliffs
Immortal in their change, and best, in the
Sad timelessness of sea and shore's
Tumultuous embrace.

 — Mary McGowan
  1956

CONTENTS

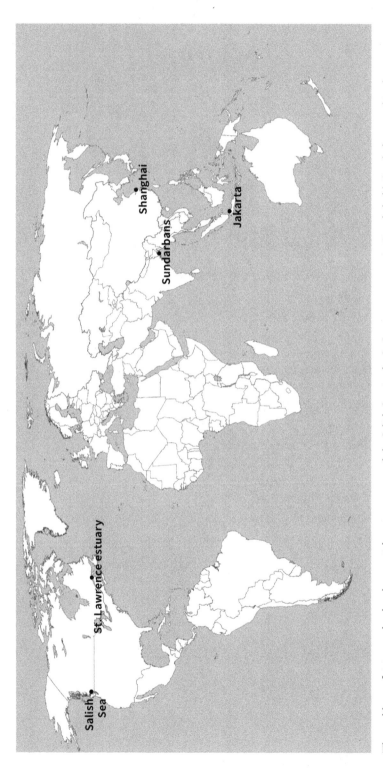

The problems of rising sea level and more violent storms around the Salish Sea, in the St. Lawrence estuary, in Bangladesh's Sundarbans, and in the great port cities of Shanghai and Jakarta illustrate what we all are up against as climate changes and the waters rise.

PREFACE

started seriously thinking about rising sea levels due to climate change a few summers ago, when I was working on a book about concrete, *Concrete: From Ancient Origins to a Problematic Future.* I'd dragged my husband along to see a newly finished cement factory on the south shore of Quebec's scenic Gaspé Peninsula. The contrasts between the landscape and the project were striking: the plant, located at a remote and nearly pristine site, would soon be sending 1,028 kilograms of carbon dioxide (CO_2) into the air for every metric ton of cement produced. Even though it was billed as the most environmentally friendly facility in North America — and the first new one in a couple of decades — when operating at full capacity it would be the biggest CO_2 producer in the province.

Beyond the perimeter of the plant, the scenery was glorious, and after I'd done my interviews, we spent a couple of days driving around the peninsula, enjoying countryside that in places reminded me strangely (or so I thought at the time) of my many trips around the Salish Sea, the arm of the Pacific Ocean that runs from north of Vancouver, British Columbia, to south of Seattle, Washington. When we turned back west toward Montreal, thoughts about the land and the danger posed to it by rising sea levels, themselves caused by climate change, crashed into my consciousness the way the waves along the St. Lawrence estuary attacked the shore.

The highway runs for about sixty kilometres along the base of cliffs that are protected by loads of concrete rip-rap, and even on this rather calm day, waves sent sprays of water across the road in places. For a good part of the rest of the way home, I mused about this, thinking of all the roads I'd driven along that would soon be prey to rising water, or perhaps were already.

The section of Interstate 5 in Washington State that runs across the Skagit River floodplain, with its colourful fields of tulips, came to mind, as did the stretch of British Columbia's Highway 99 where it crosses the Fraser River floodplain. I thought of other trips we'd taken, like one through the eastern United States around Chesapeake Bay, itself the product of river valleys inundated by past rising seas. I found myself remembering all those videos of high tides in Miami and hurricane damage in Louisiana, not to mention the news stories about Venice being submerged and islands melting away in Bangladesh, where the combined flow of the Padma, Brahmaputra, and Meghna Rivers enters the Bay of Bengal. Then there are island nations like Kiribati, the Marshall Islands, Tuvalu, and the Maldives — all of them may actually disappear beneath the waves.

Nevertheless, at the time I had great hopes that some of us would be able to lead the way to a safer, saner planet, where the danger of being drowned in your bed was infinitesimally small.

But the outlook, my friends, is not good. Barring a major reversal in trends — or some miracles — sea levels will continue to rise and the frequency of extreme weather events, ranging from torrential rains to prolonged droughts, will continue to increase. The most optimistic forecasts call for an average of 1.5°C above pre–Industrial Age values even if we drastically cut greenhouse gas emissions and, if possible, suck CO_2 out of the air.[1] There is no question about whether we are going to have to make some adjustments. The questions are these: When? Are the measures already taken anywhere near what will be required? If not, what must we do to save civilization?

This book is about how people have coped for millennia with rising sea levels, which, believe it or not, are nothing new. For a very long stretch of time in human terms, folks who lived on the margins of oceans have been

"There is nothing in the world more soft and weak than water, yet for attacking things that are hard and strong there is nothing that surpasses it, nothing that can take its place."
— Lao Tzu

challenged by the sea, as the volume of water has increased due to melting glaciers, and so they have frequently had to move. Their story twists and turns the way a shoreline does, and ends — well, we don't know yet where it will end, and that is part of its attraction.

My idea had been to start telling this story with a visit to Jakarta, the capital of Indonesia. On the northern shore of the island of Java, the city and its accompanying suburbs are home to about twenty million people. Built on a floodplain at the confluence of several rivers, it has been increasingly threatened by a combination of rising sea levels and sinking land. More about that later, but for the moment, note that when the government announced definitive plans to move the seat of government some 1,300 kilometres to the east, on another island, work was scheduled to begin in the summer of 2020.

What a great example of what so many cities are up against, I thought. I have to see what's going on there. So I nosed around to determine what was possible, even got a travel grant from my local arts funding body to research the story, and made plans to visit Jakarta and the new, unsullied capital site in the spring of 2020.

Ha! The gods had other ideas. In fact, I've often thanked my lucky stars that I didn't go then, because I could have been stranded there as Covid-19 washed across the world. By late March 2020 it became obvious that my project was going to have to be reworked. Yet the more I learned about the history of the oceans and human civilization, the more I realized that Jakarta and the Indonesian archipelago are the perfect place to start thinking about a strategic retreat from the edges of the seas.

So that's where I'll begin, going into humanity's deep history, and then looking at how rising seas have found their way into our collective mythology. After that, we'll take a look at the many ways people have tried to confront rising waters, reclaim sunken lands, or even dredge up new territory.[2] Along the way there will be several stories told and much talk about the importance of memories as the treasure chest of civilizations.

The Netherlands will get a lot of attention, both because of its pioneering work in making the delta of the Rhine the most productive agricultural land in the world and because of its contributions to shaping other deltas around the world, including the one on which Jakarta sits. Four other places will get major attention: Shanghai and the Yangtze delta in China; the Sundarbans and Dhaka in Bangladesh; the Salish Sea of the Pacific Northwest; and the St. Lawrence estuary. The first because China's enormous progress in the last forty years shows the stakes in the battle, the second because it points the way to a lower-tech approach, the third because it suggests what may happen in many places in North America, and the fourth because it illustrates some of the solutions to the problems of rising sea levels (and besides, I know it well).

Large parts of all of them could be underwater by the turn of the next century, if current projections are correct. They will be affected not only by a slow but steady increase in mean sea levels, but also by the same forces that are melting glaciers and engendering violent and invasive storms. These raise the danger of sea-level rise exponentially, as well as increasing the risk of floods from torrential rains upstream.

In a few places in the book, you'll find musical interludes. They're there to draw attention to the accomplishments of civilization and to stress how music and stories and art can help us get through the worst of times. And to be sure, very difficult times lie ahead. Even if we are able to control greenhouse gas emissions and limit temperature rise over the next few decades, the forces we've unleashed already are like the sorrows that escaped from

4

Pandora's box in Greek mythology and that couldn't be brought under control. The steady rise of temperatures may be stopped, but higher sea levels are not going to go away, because once the glaciers have melted, the water will remain.

The book will end with some thoughts on where we go from here. Amazingly, it looks as if both low-tech solutions, like just letting the waves roll in, and high-tech ones, like massive switching to nuclear-generated electricity, have merit. Over all will stretch the rainbow that the Hebrew god offered as a pledge following one mythic flood. The image gives hope as we face the rising waters. Its continuing power can be seen in the way that children in my part of the world latched on to it in the first, darkest days of the Covid-19 pandemic. Their motto was *Ça va bien aller* — it's going to be all right. Let us hope that will be the case in this other, potentially much worse global crisis.

PART 1

RISING WATERS

MUSICAL INTERLUDE

Maybe the best place to start this investigation is by the water, specifically on a dock by the bay. Doesn't matter which bay, really, but one that comes to mind immediately is the one that American soul singer Otis Redding liked back in the 1960s. In his best-known song, "(Sittin' On) The Dock of the Bay," Redding sings about watching the tides of the San Francisco Bay roll away. Despite the pulsing beat, the song is mournful, because, it says, nothing is going to change, everything will remain the same. It's not clear what Redding is lamenting: love gone wrong or good fortune that passed him by or social justice that never seems to come. No matter, several decades later, the song's ending resonates resoundingly in another register, that of climate change and rising sea levels. Redding sings that he's wasting time, which is exactly what we must not do if we want to come to grips with rising waters. Keep that in mind as we roll onward through this book.

For a great rendition of the song, watch the Playing for Change video featuring musicians around the world on the Playing for Change website or on their YouTube channel.[1] Wouldn't it be great if we could all work together the way they do.

1

THE SINKING CITY

August 26, 2019

Indonesia announces site of capital city to replace sinking Jakarta

Indonesia has announced plans to move its capital from the climate-threatened megalopolis of Jakarta to the sparsely populated island of Borneo, which is home to some of the world's greatest tropical rainforests.

President Joko Widodo said the move was necessary because the burden on Jakarta was "too heavy," but environmentalists said the $33bn relocation needed to be carefully handled or it would result in fleeing one ecological disaster only to create another.[2]

≈

January 15, 2020

Two islands vanish, four more may soon sink, Walhi blames environmental problems

Two small islands in South Sumatra have disappeared as a result of rising sea levels driven by climate change, while four other islands are already on the brink of vanishing, the Indonesian Forum for the Environment (Walhi) has claimed.

The province's Betet Island and Gundul Island — which technically fell under the administration of Banyuasin regency — have submerged, currently sitting 1 meter and 3 meters below sea level, respectively, according to Walhi data.[3]

~

J akarta lies on an embayment on the northern shore of Java, where thirteen rivers come together to enter the ocean. As a result of the geography and climate of the area — storm tides and intense rain are common — the low-lying land on which the city is located regularly floods. Despite this, the site has been used as a port since the early years of the Common Era.

In the late sixteenth century, Dutch merchants began to use the safe harbour there as a staging ground for their spice trading in that part of the world. The Dutch, who by that time were well launched into their effort to protect their own fields and cities back home, introduced a clever system of canals and holding ponds that for a while served to tame the waters. But hundreds of years have passed, and as the twenty-first century has advanced, it has become clear that the combination of rising seas, more frequent torrential rains, and the pumping of groundwater is resulting in a sinking city that is less and less suitable as a site for a government centre.

In August 2019, exactly four hundred years after the Dutch established their trading post at what is now Jakarta, Indonesian president Joko Widodo decreed that all would be better if the government were moved to a different island, a couple of hours away by air in the East Kalimantan province on the island of Borneo. In contrast to Jakarta, which is on Indonesia's westernmost island, the new capital, to be called Nusantara, will be more or less equidistant from the eastern and western extremities of the island nation. It also would be on higher ground and in a region less often rocked by seismic chills and thrills. Some fifteen or sixteen million of the people living in greater Jakarta would stay where they were, however; only the government officials, their families, and the services they would rely on would move house.

There were ceremonial announcements and extended videos of the site. The multinational consulting firm McKinsey was engaged to do strategic planning. Tony Blair, the former British prime minister, signed on as an important adviser. The government boasted that investors were lining up to help make the new city a technological wonder set in an environmentally friendly paradise.

At the time the announcement was made, the country was doing very well economically and boasted of cutting its poverty rate in half over the previous twenty years, from a high of around 20 percent in 1999 to around 10 percent in 2019. Indeed, according to the World Bank, it had reached "upper middle-income status" and had become the world's tenth-largest economy in terms of purchasing power. This relative prosperity has not been without its downside: Indonesia is the tenth largest greenhouse gas–emitting nation in the world, although on a per capita basis, it is one of the least polluting.[4] In fact, Indonesia contributes greatly to the problem of rising seas and climate change that it must deal with, since the nation relies heavily on coal-fired electricity-generating plants. It also is the world's biggest exporter of thermal coal, and for extensive periods over the last thirty years has allowed, and in some cases encouraged, the burning of its tropical forests to replant in oil palms.

Fascinating, I thought. I wanted to see the situation first-hand, and I got as far as starting to study Bahasa Indonesia (the official language) and making tentative reservations when the Covid-19 pandemic arrived. It didn't help that Indonesia was on the back of the curve, with Covid-19 really taking hold in late April 2020. Things got progressively worse afterward, due to a number of missteps that I won't go into here. Suffice to say that one

Indonesian public health official told the *New York Times* that the country initially had a health minister who thought the disease could be stopped by prayer.[5] Nevertheless, in the meantime I started to work in earnest on the book and to muse about what would happen when changes dictated by a global pandemic encountered changes that were going to become necessary because of global climate disaster.

What I didn't know when I started work was that the islands of Indonesia offer a great vantage point from which to consider just how humans have interacted with sea and savannah for thousands of years.

∼

Before we go any further, it might be wise to say a word here about the difference between rising sea levels and tides. The latter are caused by the gravitational pull of the sun and the moon on the waters of the Earth. For as long as there have been oceans, the water in them has bulged toward the heavenly bodies as the Earth rotates. This creates an oscillation that carries water up and down a coastal margin twice a day. The highest tides come at the full moon and the new moon because that's when the Earth, moon, and sun are aligned. But there are times when tides are higher than the usual high tides, particularly around the equinoxes — that is, the start of spring and fall. The cosmos are aligned, literally, so that there will be maximum swing in the oceans' waters during the twelve-hour period of the tides. In some places, these are called king tides, but they occur everywhere and give a glimpse of what rising sea levels will bring on a permanent basis. If they are combined with a weather "event," watch out!

∼

It is important to realize that the issue of rising sea levels we are dealing with today has been encountered before; many times in the past, humans have been faced with waters that lap higher and higher with each season. In fact, even before the first group of humans successfully walked out of Africa, there had been numerous encounters with rising sea levels.

Another thing to remember is that a rise in sea level is not like the filling of your bathtub. If it were, accommodation would be easier and could be accomplished in slow steps that might give people the time to figure out the best way to cope. No, while a barely noticeable rise in water levels continues in the background, the great damage that is done comes more suddenly, when storms whip up waves that magnify high tides, engulfing houses and encampments, eroding cliffs, breaching sandbars, and pushing salt water up rivers and into groundwater.

At times in the past, a fight was put up: seawalls were constructed, coastal waters were dredged, magic was invoked. Frequently, it became clear that the community had to move.

Yes, *had to move*. Again and again people must have decided to do just that. We'll never know what kind of discussions a group might have had, whether the old ones were ignored as they complained (as the old so often do) that the world was going to hell in a handbasket, or whether their memories and wisdom were valued and used to help guide the group's actions. Nor do we know if scouts were sent out to see what country might be promising elsewhere. Or if the collapse of a cliff to the waves and winds of a big storm led to flight in a matter of days or, worse, sudden death.

What we do know is that people have had to cope with rising sea levels for a very long time.

This is a story of how they did that. It is a story that should give hope, because it tells of the resilience of people, of their collective wisdom, of their spirit. It is also a cautionary tale: the effects of rising sea levels aren't going to be the same on everyone. Even people who live in more or less the same place will be affected differently, and that has implications for what we should do and how we should do it.

Jakarta is a case in point. The crowded housing and the filthy canals of the old city will remain when the capital moves; the people who will be running the show will be the ones who relocate. Removing a couple million people from Jakarta will not reduce the size of the city in any significant way. The problems of a sinking city and rising sea levels will be far from solved.

Jakarta grew tremendously in the twentieth century: in 1900 its population was about 115,000, but by the end of the century, the size of the core city and its satellites were orders of magnitude greater.[6] Its more central part is served by a modern infrastructure system — subways, trains, huge shopping centres, highways. But the Ciliwung, the major river in the delta, is one of the most polluted rivers in the world, and less than 20 percent of the city's residents have piped water.[7] The poorest must buy it by the jerry can, while the richest pump it from the aquifer underlying the city. As a result of subsidence caused by the taking of water from underneath the city and a rising sea level due to climate change, 240 square kilometres of the city is now actually below sea level, even when the weather is good and tides are not pushed by winds. In addition, rainfall there averages 1.79 metres a year, about the height of the average North American male; in short, the potential for flooding is immense.

It was not always like this, though. Indeed, Java wasn't even an island when the first people like us arrived.

To understand that, we have to go back — way back.

Several years ago, when I was writing my book *Road Through Time: The Story of Humanity on the Move*, I tried to imagine what it would have been like for that first group of modern humans to leave Africa.[8] The idea was to try to figure out how people have moved around the world since modern humans evolved. My starting point was a consideration of roads, which seemed to me, after a couple of decades of thinking about cities and science, to be the most enduring creations humans have left behind. But if you go back far enough, you realize that before there were roads, or even trails, people must have been just walking toward places that they knew nothing about, travelling without a map, knowing only that the things they saw from a distance promised food and fresh water. Or perhaps adventure! The story I imagined for the book proposed that the first crossing of the short distance between the Horn of Africa and the southern Arabian peninsula might have been an accident. Today the Bab el-Mandeb Strait — thought by most researchers to be the gateway east for early modern humans — is only twenty-six kilometres wide, but during a time when sea levels were much lower than they are now, the distance would have been considerably less. Perhaps some adventurous fishing party or mollusc gathers rafted across. I built a whole scenario around that idea.

Viewed from the vantage of several years later, as paleontological research has advanced, it looks like I had some of the story right, but not all. Now it appears that other groups of modern humans strayed from their East African homeland as early as three hundred thousand years ago,[9] while others tried to settle greener fields along the southern and eastern Mediterranean about two hundred thousand years after that.[10] Human bones dated to more than one hundred thousand years ago were found in what's now Israel in the early twentieth century, but this migration appears to have been a dead end.[11] Recent analysis of their DNA suggests that nothing from this small population remains in the genes of living persons. But a large part of the story I recounted earlier still works: it seems clear that the "founder" population that left Africa took advantage of lower sea levels to make the crossing.[12]

The problem with trying to document this, of course, is that a good part of the evidence attesting to this version of human history must now be underwater. Had these pioneers explored by keeping to the margins of the seas, their campsites and temporary settlements would have been covered by rising seas over the millennia. That is why one of the most dynamic branches of archeology today is that which aims to study what lies beneath the waves; we'll return to discoveries from this field later.

Sea levels depend on many things, even though the amount of water on Earth in its various forms is more or less constant.[13] The key phrase here is *in its various forms*. For long periods in the Earth's history, an immense amount of water has been locked up in ice. The reasons for this are cosmic, it seems. The Earth wobbles a bit on its axis, and its path around the sun changes in predictable ways. Taken together, these movements result in differences in the amount of sunlight falling on the Earth, which produce dramatic results. Not only was there a period about two billion years ago when the whole planet was frozen, but there have been four other major ice ages since then. We're about three million years into the most recent ice age, in fact, although that may seem difficult to square with the present-day concern about climate change and global warming. Our current rising temperatures are an anomaly, caused by human activity, but that's getting ahead of the story. What matters here is that the long-long-term forecast has been for cold, although within that framework there have been several ups and downs — changes that have had immense consequences for sojourning humans.

How do we know just what the weather record has been since the beginning of human history? Scientists have come up with several very clever ways to ferret out clues. Ice cores taken from the glaciers and the icefields of Greenland, the Arctic, and the Antarctic give insight into the situation over the last 450,000 years, a swath of time that encompasses all of the history of anatomically modern humans, as well as much of that of our cousin hominins, like *Homo erectus*, Neanderthals, and Denisovans.[14] The cores clearly show temperature changes on an annual basis, giving both a good measure of time elapsed and what the weather was like that year, much the way growth rings on trees are a graphic record of the weather during the tree's life. In addition, air bubbles trapped in the ice suggest how the atmosphere has changed, with the amount of CO_2 increasing rapidly after about 1800 not long after the start of the Industrial Revolution.

These ice cores show hundreds of thousands of years of deep-freeze, with five periods of relative warming — called interglacial periods — when average temperatures approached what we're experiencing today. The track they make on a graph looks like a cross-section of an ancient jawbone that has lost some teeth. Strikingly, there are abrupt peaks where average temperature rises by twenty degrees or more, sometimes within a decade, followed by a slower, more gradual descent to bone-chilling average temperatures. The result of these fluctuations in ice is corroborated by research done by taking cores of the ocean floor and studying the growth of coral reefs.

If we look at these studies and superimpose on them the timeline of human history, it appears that during much of the time we spent evolving on the African savannahs, temperatures were considerably colder than they are now. The result was that far more water was bound up in ice than is the case presently. As a consequence, sea levels were considerably lower than those of today for much of the time. Indeed, it looks as if they fluctuated from a high of 4 to 5 metres above their current levels at the beginning of the last big ice age, three million years ago, to a low of around 120 to 150 metres below current levels at the deepest point of the most recent deep-freeze, about twenty thousand years ago.[15]

These changes, often called eustatic changes, affected the entire world.[16] The tracks of great weather systems, like the monsoons that sweep out of the equatorial regions, shifted with these changes, so despite the fact that there was less liquid water to be carried in the atmosphere, conditions for millennia were much wetter in parts of Africa and in southwest Asia. This meant that not only were conditions great for hunters and gathers on African savannahs, with water and grasses and game, but also, once early modern humans moved out of Africa, they found similar green pastures, plus abundant fish and shellfish along the coast and the rivers they encountered.

It should be noted that *Homo sapiens* weren't the only hominins around in these new frontiers for humans. Neanderthals had ranged widely throughout Eurasia for as long as three hundred thousand years, while another group that has recently been identified, the Denisovans, appear to have been roughly contemporary with Neanderthals and thrived in what are now parts of Asia. Before them, *Homo erectus* roamed for more than a million years from Africa to as far north as France and as far east as Java, the island where today's Jakarta is located.[17]

These cousins of ours were far from being primitive brutes. Their long history and wide range suggest they were a very successful species, with enough social organization to make plans and to cross relatively wide stretches of water in craft that probably resembled large rafts.[18] That sounds nearly impossible, but there is no doubt that they made crossings between islands. Their fossils are the only ones of large mammals present on islands east of the Wallace Line (about which more later) except for those of stegodonts, large elephant-like animals that once roamed much of Asia. How the stegodonts got there is less of a mystery: it is generally believed now that they swam to the islands of Java and Flores. Their modern elephant cousins have been observed swimming for up to fifty kilometres in Africa, while in the Indian Ocean and Indonesia, elephants were used until recently to tow wood and other material between islands.[19]

Be that as it may, while genes identified as coming from Neanderthals and Denisovans show up in most folks today, testifying to the fact that early people like us loved or lusted after some of our hominin cousins, DNA from *Homo erectus* hasn't yet been sequenced, so we don't know if any of us carry genes specific to them.[20]

On the other hand, nearly all current evidence points to only one successful founder population of anatomically modern humans who left Africa. Some estimates based on mitochondrial DNA — those bits of genetic information that control the packets of energy powering all of us and are handed down from mothers to their children — suggest that this group might have been as small as one hundred, but probably not as large as several thousands. Nevertheless, once these folks had made it onto the next continent to the east, they prospered in what might be called the Great Expansion.

The possibility that modern humans travelled quickly east during the Great Expansion has recently been reinforced by a re-evaluation of fossil evidence and artifacts from Sumatra, the island just to the northwest of Java.[21] A human tooth found in the 1880s at Lida Ajer cave in Sumatra's tropical rainforests has recently been dated to between 63,000 to 73,000 years ago. That's much earlier than the dates for the previously oldest evidence of people like us in this part of the world, found in Madjedbebe, a rock shelter in northern Australia. There, a complex of artifacts and fossils of what seems to be leftovers of plant foods prepared by early modern humans has been dated to at least 53,000 to 63,000. There is no doubt that *Homo sapiens* successfully colonized Australia tens of thousands of years ago by using some kind of watercraft, even though when Europeans arrived the First Nations populations did not have ocean-going vessels. This anomaly led some to postulate in the early twentieth century that, rather than coming out of Africa, First Nations Australians were the children of a separate line of evolution.

That hypothesis has conclusively been proved wrong — the DNA research shows that First Nations Australians are related to all of the rest of us who started out in Africa[22] — but the question of how they could get there so early in the history of the Great Expansion remained a puzzle for a long time. Putting the pieces together presents a picture that helps illustrate what is happening in Jakarta today, and what will happen to the whole world in the future.

A key piece of the puzzle was discovered more than 150 years ago by one of those curious British adventurers who roamed the world in the nineteenth century. Alfred Russel Wallace spent eight years travelling around the islands of the western Pacific, and he left hundreds of well-written pages describing his journeys and elaborating on the lessons that he deduced from them.[23] He was a peculiarly nineteenth-century character, a man with little formal education but with an inquiring mind and a passion for Nature.

It was a time when collecting specimens of plants, insects, and animals was extremely popular among curious and intelligent people of all social classes in the United Kingdom, Europe, and North America. Wallace's aim in his travels was to satisfy his own curiosity but also to supply specimens for the wonder and edification of others; he was successful enough at selling them that he could finance his travels with the proceeds. A keen observer of the world he was visiting, he wrote dozens of scholarly papers and sent them back to England for circulation among the scientific community.

Today Wallace is known for two things. One is his theory of natural selection and evolution of species, which he arrived at about the same time that Charles Darwin was elaborating his own world-shaking ideas. The other is his observation of the immense difference in life between the eastern and western section of the group of islands — let us call them the Ocean Isles, because not all are in Indonesia today — in which he travelled.[24]

His first major accomplishment has been discussed many other places,[25] but his second insight is perhaps less well-known to people in the Americas and Europe. What's called the Wallace Line is literally on the other side of the world from them. Furthermore, at first glance what it marks seems improbable, and it's only by chance that Wallace saw it up close.

He'd spent two years in the region, travelling around, gathering specimens, and sending home scientific articles and notes, when he decided to visit Macassar, a city on the island now called Sulawesi, but which was then called Celebes. In *The Malay Archipelago*, Volume 1: *The Land of the Orangutan and the Bird of Paradise*, he writes, "Had I been able to obtain a passage direct to that place from Singapore, I should probably never have gone near [Bali and Lombok], and should have missed some of the most important discoveries of my whole expedition [to] the East."

The vessel he was travelling on made a call on the north side of Bali, an island that has become a very popular tourist destination. It is, according to *Lonely Planet Indonesia*, worth visiting for its "impossibly green rice terraces, pounding surf, enchanting Hindu temple ceremonies, mesmerising dance performances, ribbons of beaches and truly charming people." The guide adds, "There are as many images of Bali as there are flowers on the island's ubiquitous frangipani trees."

Wallace was also impressed. Among other things, he wrote, "I had never beheld so beautiful and well cultivated a district out of Europe." However, because it presented such a blooming, cared-for landscape, he says that he didn't take as many specimens as he should have, since he assumed that intense cultivation would have affected the native flora and fauna. This error became apparent only after his boat continued on toward Lombok, just a short sail away. Of the voyage, he writes, "We enjoyed superb views of the twin volcanoes of Bali and Lombock, each about eight thousand feet high, which form magnificent objects at sunrise and sunset, when they rise out of the mists and clouds that surround their bases, glowing with the rich and changing tints of these the most charming moments in a tropical day."

This first impression of similarities soon altered after he spent several days exploring Lombok.

> In Bali we have barbets, fruit-thrushes, and woodpeckers; on passing over to Lombock these are seen no more, but we have abundance of cockatoos, honeysuckers, and brush-turkeys, which are equally unknown in Bali, or any island further west.... The strait is here fifteen miles wide, so that we may pass in two hours from one great division of the earth to another, differing as essentially in their animal life as Europe does from America. If we travel from Java or Borneo to Celebes or the Moluccas, the difference is still more striking. In the first, the forests abound in monkeys of many kinds, wild cats, deer, civets, and otters, and numerous varieties of squirrels are constantly met with. In the latter none of these occur; but the prehensile-tailed Cuscus

is almost the only terrestrial mammal.... The birds which are most abundant in the Western Islands are woodpeckers, barbets, trogons, fruit-thrushes, and leaf-thrushes; they are seen daily, and form the great ornithological features of the country. In the Eastern Islands these are absolutely unknown, honeysuckers and small lories being the most common birds, so that the naturalist feels himself in a new world, and can hardly realize that he has passed from the one region to the other in a few days, without ever being out of sight of land.

He concludes: "The inference that we must draw from these facts is, undoubtedly, that the whole of the islands eastward beyond Java and Borneo do essentially form a part of a former Australian or Pacific continent, although some of them may never have been actually joined to it," while those to the west of what has become known as the Wallace Line were, at one point, part of the great southeast Asian land mass.

Wallace knew well that the sea out of which the western islands rise is generally not very deep. Nor is the sea between the eastern islands and Australia. He also knew that in a few places, like the strait between Bali and Lombok, the seabed plunges precipitously. This means that even though the distance that separates islands on either side of the trench isn't wide, getting across the gap would require some kind of sea journey, something deemed highly unlikely before early modern humans had figured out how to make boats. (He notes that wild pigs were found on the eastern islands and asserts that they had been transported there in recent times by people.)

All right, Wallace reasoned, something must have happened to account for the big differences. Noting the widespread volcanic action in the region, he suggested that long ago everything was connected, but that the ground subsided because of seismic activity, with the shallow seas a result.

When he published that idea in 1869, it was part of the great scientific reevaluation of the world that was underway in the middle of the nineteenth century. Beginning with pioneer works on geology published in the 1830s, the age of the Earth was pushed back from a mere six thousand years (as calculated by a clergyman who took the Bible seriously[26]) to some millions of years, in which small changes over time could add up to massive differences. This gradual approach stood in marked contrast to catastrophism, which took the biblical Flood as the origin of all strange, hard-to-explain things found as men (and some women) dug up fossils and speculated about the origin of mountains and the geology of plains. Given Wallace's keen interest in trends in scientific thinking, it is not surprising that the idea of gradual subsidence occurred to him as the mechanism that resulted in the vast differences on either side of the strait between Bali and Lombok.

Wallace certainly was right about the intimate connection between seismic activity and the arc of the Ocean Isles, including present-day Indonesia: the island arc's origin lies in the great waltz of continents across the Earth's surface. As he observed, the western part of this conglomeration of islands is located in very shallow seas, so shallow that when sea levels were lower, they clearly would have been connected. To the east, Australia, New Guinea, and a wealth of other islands would have also formed one land mass.

But rather than subsidence of the sea, the deep history of the islands, we've come to learn, is wrapped up in the continuing movement of the huge sections of the Earth's crust now called tectonic plates. The reasons for their movement lie in the way they rest on a layer of semi-molten rock called the mantle, which is made up of rock superheated by forces inside the Earth. The mantle is not static but appears to have as many currents in it as a glass beaker of boiling water has columns of rising bubbles. These currents raft sections of the crust around the world, creating what can only be called havoc when they come together or when they part.

The Ocean Isles were formed as four of these plates — two from continents and two from the ocean — jostled together. One of the oceanic plates was forced under the Eurasian continental plate, which resulted in the formation of a chain of islands that is among the most seismically active regions on Earth. The islands of Sumatra, Java, Bali, and Nusa Tenggara have their roots in volcanoes that sprang forth when the molten rock under the converging pieces of crust was forced up and out.

This collision between tectonic plates is still going on. Java itself was formed only about two to three million years ago, well after the time when the ancestors of modern humans and those of chimpanzees diverged back in Africa. Today, Sumatra is cut horizontally by a still-active strike-slip fault that is about two thousand kilometres long, with cleavages that extend into the sea. One of the strongest earthquakes of recent times struck off the western coast of Sumatra on December 26, 2004.[27] Measuring 9.1 on the Richter scale, the quake caused part of the sea floor to surge upward, displacing billions of tons of seawater. This tsunami wave decimated the Sumatran coast to the east but also roared westward toward India. In all, around 225,000 persons died in the disaster, that, as we shall see, stripped away sand that had covered up the ruins of temples built when the sea level on the eastern coast of India was much lower.

The shaking and top-blowing continue. In the spring of 2020, not only were there three earthquakes in the Indonesian archipelago stronger than 4.0 on the Richter scale, including one rated at 6.7, but two volcanoes belched rocks, smoke, ashes, and lava. One of them was Krakatoa, a volcanic island between Sumatra and Java, which blew its top in 1883. It captured the imagination of the world, which learned of the explosion within hours through the recently installed worldwide network of telegraph communications.[28]

Over the centuries, events like these must have affected the beings that lived on the islands, but rising sea levels probably affected them as much or more. If the movement of tectonic plates is a sort of slow, passionate waltz punctuated by moments of intense action, then change in sea levels might be compared to the cha-cha — back, forward, back again; up, then down, then up and up and up.

When early modern humans reached Australia, sea levels around the world were about eighty to one hundred metres lower than they are now. At that time, the peninsulas and islands of Southeast Asia, the eastern Indian Ocean, and the westernmost Pacific formed two large land masses, just as Wallace suggested. Modern-day geologists have dubbed the western part that encompasses what is now Malaysia, parts of the Indochinese peninsula, the Andaman and Nicobar Islands of the Indian Ocean, and a good part of Indonesia as Sundaland. The name comes from a civilization that thousands of years later would prosper on the island of Java. The eastern land mass that consisted of Australia, Papua New Guinea, Tasmania, and assorted islands is called Sahul, a term that shows up on seventeenth-century Dutch maps of the region attached to banks — areas of shallow bottom — that were navigational hazards.

The path these human pioneers took to get to what is now Australia and New Guinea is not clear. They could have been following the shorelines, as it seems so many early modern humans did during the Great Expansion. Or they might have come down the interior of Sundaland. The region has been home to forests of tropical trees over recent millennia, but during this phase of the Great Expansion, its landscape appears to have been a mixture of grasslands and open forest, not unlike the savannah lands of East Africa where *Homo sapiens* evolved. This is a landscape where, cross-cultural studies have shown, most humans feel comfortable. Indeed, our skill in running — humans are the world's fastest long-distance runners[29] — and our preference for grassland landscapes appear to be hardwired legacies of that evolution.[30]

Forests, on the other hand, can be hard going for humans, unlike for our distant African cousins the chimpanzees and — to use an example from Indonesia — the orangutans, who spend a large portion of their time in trees. It's possible, indeed, that in Australia those first human inhabitants found just what they liked in a landscape; the continent appears to never have had the sort of jungles and forests that are found in either present-day Southeast Asia or northern Europe. (However, that human tooth found at Lida Ajer, the cave deep in the tropical forest of Sumatra, proves that some early modern humans found a way to thrive in them, too.)

Nor were the First Peoples in Australia deterred by the fact that to get from Sundaland to Sahul required crossing the seas.[31] Just what route they took and what watercraft they used is not known; as I've said, there is still controversy about when they made the crossing or crossings. Without a doubt, though, other folks living on the islands of what is now Indonesia had become adept at braving the ocean waves by forty-three thousand years ago. A wealth of fishing equipment and the bones of tuna and other fish that can't be caught close to shore have been found on Timor, one of the most eastern of Indonesia's islands (*timur*, in fact, means "east" in Bahasa Indonesia).[32] Remains of these early fishers' boats have not yet been found, however, just as the craft that initially ferried people like us to Australia remain unknown.

What is perfectly clear, however, is that the people who became today's First Nations Australians successfully colonized the continent, and then were cut off from the rest of the world for tens of thousands of years.

Why that happened is a key part of this story of how people have dealt with rising sea levels.

As we saw earlier, despite the current trend of average temperatures rising higher and higher, we are technically still in an ice age, one of several that have profoundly affected the Earth's climate, with direct repercussions on the amount of liquid water in the oceans. The take-away from this has to be that our ancestors came up against sometimes rapid fluctuations, changes that were perceptible within the lifetime of an ordinary person. In the next chapter, we'll look at several instances of these sudden changes and the artifacts the people who lived through them left behind.

But first, we should look more closely at Jakarta, which in many respects is the poster child for cities coming to grips with climate change.

~

Jakarta is both a modern port and an ancient one.

When does a city's history begin? In some cases, the answer is clear; there will be no doubt when the new Indonesian capital is open for business. Decrees are made, projects undertaken, buildings put up, official histories begun. We'll talk quite a bit about such cities when we get around to considering what could and perhaps should be done to protect our civilizations from rising sea levels. But most of the world's cities were not planned from the beginning, although they may have undergone some sort of planned transformation later by necessity or official whim. A substantial proportion of the world's cities and villages and settlements have origins that are lost in the mists of time.

Jakarta is no exception. Among the questions whose answers are obscure is this: Where did the ancestors of the people who now live in Jakarta come from originally? Clues can be found in the history of the languages spoken there.

Bahasa Indonesia is one of more than 1,200 Austronesian languages. They are spoken by about 385 million people in a wide swath of the world that extends from parts of the Southeast Asian peninsula south through Malaysia and Indonesia, then east as far as Easter Island, and west to Madagascar. Just as Latin became Italian, French, Spanish, and Portuguese over the centuries after the fall of Rome, many separate languages evolved in the Ocean Isles. One version — Trade Malay — was spoken much the way Swahili is spoken today in Central and East Africa. Both are lingua francas, auxiliary languages that contain elements of different languages, used so that their speakers can communicate with each other. When Indonesia began to agitate for independence, patriots chose to use Malay as a unifying force for the nation they were hoping to create. Their motto was *Indonesia — satu bangsa, satu bahasa, satu tana* (Indonesia — one people, one language, one fatherland),[33] and Bahasa Indonesia was chosen as the new country's official language when it became independent from the Netherlands in 1947.

Linguistic scholars now think that the original Austronesian language originally developed in Taiwan during a period of low sea levels, when there was easy passage from mainland China by foot or by short hops by boat. People moving to what is now an island brought with them a culture kit of rice, pottery techniques, and other things that allowed them to flourish.[34]

Then these early Austronesian speakers moved south into new territories, colonizing part of the Indochinese peninsula, Malaysia, and the great islands of what is now Indonesia. Along the way they met societies of hunters and gatherers who had arrived in earlier migrations, most of whom have now been pushed to the margins of society or been absorbed into the larger population. It should be noted, though, that these earlier groups were not without their resources: the oldest representative art in the world comes from one site on the Indonesian island now called Sulawesi. Deep within a cave in the jungle, wall paintings of rare beauty depicting animals and people plus constellations of the imprints of hands have been dated to between 35,000 and 39,000 BCE.[35]

Our knowledge of the people who lived on Java and its neighbouring islands is much clearer when we approach the present. From around the fourth century BCE to the first century CE or later, the floodplain of the Ciliwung River appears to have been the site of the Buni culture, characterized by a particular kind of pottery.[36] A more definite date can be given to one of the earliest written relics extant, the Tugu inscription from the fifth century CE. The monument stone, using an early Sanskrit script, records the exploit of Purnawarman, king of Tarumanagara from 395 to 435 CE, who built a canal designed both for irrigation and flood control; obviously the particular problems of Jakarta's location were recognized 1,500 years ago.[37] Tellingly, the capital for another dynasty was at Bogor, which is on the Ciliwung but upstream in the hills and south of present-day Jakarta. The location is both cooler and less prone to floods even today.

But the strategic importance of the site of the future megacity grew as trade blossomed. The Sunda Strait, which separates Java from the island of Sumatra to the northwest, is one of the few places where ships can cross from the Indian Ocean into the Pacific Ocean and then beat north to the riches of Southeast Asia, China, and Japan. The Ciliwung's floodplain lies around the corner from the strait, so to speak; that is, somewhat to the east. Although currents are strong and the strait itself is twenty-five kilometres wide at its narrowest, it was used from about the first century CE by Arab and Indian traders. It became increasingly important after European voyagers began to go south around the tip of Africa to get to the riches of the East.

Spices were key items in this trade, and Java became a major transfer point. Nutmeg, cinnamon, and cloves were native in the Banda Islands, a group of small isles to the east of Java. Sailing ships full of the coveted spices could sail with the monsoon winds from east to west during part of the year. The spices would be unloaded and traded for rice that grew abundantly on Java. Then, when the wind shifted with the arrival of the other monsoon, the ships would sail home, and the treasure would be sent to the wider world on ships sailing from ports in China or on the Indian subcontinent. So valuable was their cargo that, in the seventeenth century, during the settlement of a war half a world away, the Dutch gave up Manhattan to the British in exchange for one of the Spice Islands, Banda Rhun.

The traders carried with them their culture, as well as goods prized by the local rulers. Early in the Common Era they also introduced two of the world's most important belief systems, Hinduism and Buddhism. The writing scripts in which their holy texts were written were adopted, and their followers constructed magnificent monuments and places of worship. The earliest ones date from the third century CE, and by the ninth century, both belief systems prompted some of the largest and most elaborate temple complexes anywhere in the world. Java's oldest stone structures are eight small Hindu temples dating from around the seventh century on the Dieng Plateau, the marshy floor of a volcano's caldera about 350 kilometres from Jakarta.[38]

The Prambanan Temple Compounds, about 435 kilometres south of Jakarta, whose construction started about two hundred years earlier, is much larger, containing 508 temples. Along with shrines dedicated to Hindu worship, it also contains the Sewu complex, which is Indonesia's largest Buddhist centre. As its UNESCO World Heritage citation says, "Prambanan Temple Compounds represents not only an architectural and cultural treasure, but also a standing proof of past religious peaceful cohabitation."[39] More evidence of this open approach is Borobudur, the world's largest Buddhist temple, which was built during the same period when both Hinduism and Buddhism were practised side by side.[40]

Incidentally, some stories told about these past kingdoms bear a striking resemblance to the deep history of the islands. Sir Stamford Raffles, who was

British lieutenant governor of Java from 1811 to 1814, wrote the following in his two-volume history of Java, dedicated to George, the Prince Regent:

> Amongst the various traditions regarding the manner in which Java and the Eastern Islands were originally peopled, and the source whence its population proceeded, it has been related, that the first inhabitants came in vessels from the Red Sea (*Láut Míra*), and that, in their passage, they coasted along the shores of Hindustan; that peninsula then forming an unbroken continent with the land in the Indian Archipelago, from which it is now so widely separated, and which, according to the tradition, has since been divided into so many distinct islands, by some convulsions of nature or revolution of the elements.[41]

Echoes of the geological past in stories told from generation to generation? Possibly. As we'll see in the next chapter, oral tradition can shed much light on our experience in the long-ago past. Be that as it may, not long after their period of greatest flowering both Hinduism and Buddhism were swept aside on Java and elsewhere in the Ocean Isles by the religion brought by Arabs, the master navigators who taught so much to the sailors from India and China. Within a couple of hundred years, both Hinduism and Buddhism were marginalized, to be replaced by a brand of Islam carried by these traders. This was part of the remarkable rise of Islam across continents, beginning in the seventh century and continuing, many would say, today.

Southeast Asia and the Ocean Isles were not the scene of militant Muslim aggression, although once established, some Muslim rulers, including those on the northern shore of Java, battled with Hindu kingdoms that ruled from highlands in the interior. But the brand of Islam that took hold in the Ocean Isles was furthered by trade and intermarriage and, to speak in terms of belief, a direct connection with the deity that other faiths did not offer. Its triumph is clear today: in the 2020 census, some 86.7 percent of all Indonesians professed Islam, while 10.72 percent said they were Christians.

As for Hinduism and Buddhism, once such powerful spiritual forces, each had less that 2 percent of the population.[42]

Most of the Christians are the spiritual descendants of the Portuguese who were the first Europeans to claim a foothold in the Ocean Isles. These intrepid navigators (who nevertheless owed much to the Arab captains who taught them about travelling on monsoon winds) wanted to wrest control of the trade in spices and other riches from the Venetians, who had long been Europe's middlemen.

Christopher Columbus, who learned his navigation on Portuguese ships, was a Genoese captain who, as most North American schoolchildren know, sailed west under the Spanish flag in hopes of finding the Orient (as Europeans called East Asia at the time) by going that way. The Portuguese had a different strategy: they sailed south into the Atlantic, veering wide to the west to pick up prevailing winds that would swing them around the tip of Africa. After that, they coasted up the eastern shore of that continent until they could catch the monsoons and sail across to India. Vasco da Gama arrived for the first time on the Malabar Coast of India, the home of black pepper, in 1499 and returned again in 1502. By 1509 one of his confreres, Afonso de Albuquerque, had continued east, and in 1511, he captured the port of Malacca on the Malacca Strait, which, like the Sunda Strait, allows passage across the arc of the Ocean Isles toward the treasures of the Spice Islands, China, and Japan.

At the time, conflict between various kingdoms on Java was raging, just as it was in Europe, where many small states frequently fought each other. The Mataram Sultanate controlled much of the interior and east of Java, while the Kingdom of Bantam, or Banten, held the western end. The harbour at Bantam was the major port for trading and transshipping spices, as it was located practically on the Sunda Strait. That changed, however, when the river silted up, leaving the previously small-scale port at the mouth of the Ciliwung River as the best protected anchorage near the strait. In 1521 the king of the Sunda empire agreed to give trading rights to the Portuguese if they would build a fort there; in time it became Jakarta. Echoes of the fort can be seen in the section of the modern city called, like the previous settlement, Sunda Kelapa. There the Portuguese erected a *padrão*, a stone pillar like those the Portuguese put up elsewhere during their explorations.[43]

Portuguese influence declined markedly in the seventeenth century, a result both of local uprisings and of the rise of the Dutch East India Company in the region. Eventually Portuguese holdings in the Ocean Isles were reduced to part of the island of Timor, which was a Portuguese colony between 1702 and 1975.[44] Nevertheless, this colonial experience marked the Ocean Isles out of proportion to the length of time the Portuguese prevailed.

One way to consider the extent of a colonizer's influence is to examine the countries that speak its language. Any list of the world's most spoken languages depends greatly on who's doing the counting and whether they're counting only native speakers. In terms of native speakers, Mandarin comes at the top, but three languages of European colonizers are right up there: Spanish, English, and Portuguese. The shadow of this past colonialism still falls over much of the world. But Dutch isn't there, even though Dutch rule in the Ocean Isles lasted for nearly 350 years.

There are a lot of words for ordinary things in Bahasa Indonesia that have their roots in Portuguese — *jendela* (window) is much like the Portuguese *janela*; *sepatu* ("shoe") is like *sapato* in Portuguese; and *kemeja* (shirt) is not far from *camisa*. The list for Dutch borrowings tends to run to administrative words like *perboden* (forbidden, or *verboden* in Dutch) and *kantor* (office; *kantoor* in Dutch).[45] The reasons for this go back to first contact with the Dutch and the Portuguese. Those in charge of the Dutch trading contingent were single minded and frequently brutal in their approach to the native population. Few Dutch women accompanied their men, and liaisons between local women and Dutch men, who expected to return to the Netherlands, were at first rarely formally recognized. Many of the Portuguese in the Ocean Isles, however, established relatively long-lasting unions with local women, and Portuguese became the language of the household.[46]

Yet the Dutch undeniably left their mark in Indonesia, including on the layout of Jakarta, which they called Batavia and made the centre of their operations in 1619. They named their new city after the Batavi people, who lived on the delta formed by the Rhine and the Meuse Rivers and about whom we'll have much more to say later. In relatively short order after they took control of Java, the Dutch set out to tame the river landscape, just as folks back home were doing. One of the earliest images of Batavia is a map that shows the parallel breakwaters that extend into the sea, the wall around the city proper, and canals that run through the various districts and the fortress.

In 1835 John Crawfurd, a British colonial official, wrote about the town in his *Descriptive Dictionary of the Indian Islands and Adjacent Countries:*[47]

> [It] is situated on the shore of a bay, some 60 miles wide, but of no considerable depth.... The land on which it stands is little above the level of the sea, and consists of a recent alluvial formation.... The new town, originally suburbs of the old, lies inland from it, and is generally 30 feet above the sea level. Through both, there runs a river of no great size, but with a rapid current, having its source in the mountains of the interior, at the distance of some 50 miles. The native name of this stream [means] "perplexed river."

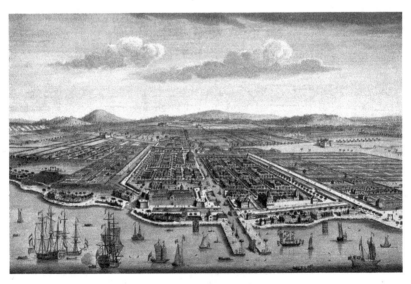

Batavia in 1754: an idealized view of a Dutch planned city.

For years after Batavia's foundation, the climate was not remarked for unhealthiness, but the European-built town soon acquired a proverbial reputation for insalubrity. The river, which ran through the town in many canals, lost its current and dropped its sediment, leaving stagnant water that was ripe for generations of malaria.

In her article "Dutch Batavia: Exposing the Hierarchy of the Dutch Colonial City," Marsely Kehoe notes that the Dutch East India Company was attracted by the familiarity of the low-lying land around the port, which led them to "continue the Dutch fight against water in the colonies, though not as successfully as at home." In the Netherlands, canals are regularly flushed out, with their water going into "the larger body of water." But at Batavia, the canals "became less functional over time because of irregular water flow from inland, leading to canals with shallow, stagnant water that were unable to flush the city's sewage into the sea." She goes on: "The Dutch designers mistakenly presumed their model would function throughout the world."

Today that statement seems a bit ironic, since the Dutch were and are pioneers in saving lands from the sea and promise to provide models and techniques going forward. But before we get to that, let us look at why floodplains and river deltas are so attractive, not only to the Dutch and the Indonesians, but all round the world, and have been from the beginning of recorded history, or even before.

~

Deltas are formed when a river, or rivers, rising in higher country flows to the sea, usually carrying a very large load of sand, soil, and stones that the water has liberated from the upstream rock. In effect, they are engaged in a slow dance with the sea. Sediments deposited by the river increase the delta, until the sea and its storms attack; think of a tango dancer pushing his partner backward until she almost loses her balance. Or until she actually does — and seawater rushes in, transforming some or most of the delta into a salt marsh.

One of the ditties that children of my generation sang in California's public schools was the English version of an old Czech folk song. Possibly one of the reasons it made such an impact on me was that it was so out of phase with the climate of the place where I lived. I don't remember all the verses, but the refrain comes to me every springtime as the snow around Montreal melts: "Streamlets down mountains flow / Pure from the winter's snow / Joining, they swiftly go / Singing of life so free." There sure weren't streams like that in San Diego; by the time rivers made it to the coast where we lived, they were muddy with a load of dirt.

The San Diego River's delta stretched out westward from the valley it had carved in the bedrock. By the time I was a child, the delta wasn't growing because the river was contained by dams and canals upstream, where the sediment dropped out. The river itself flowed sluggishly, except in rare years when storms dropped inches of rains over Southern California. Then what was called Mission Valley flooded, leaving a coating of sediment over everything.

This kind of flooding has many advantages, and it is no accident that some of the earliest and greatest civilizations depended on just such flooding or developed on land where flooding had, in the past, been frequent.

The Nile and the Egyptians, the Tigris and Euphrates and the Mesopotamians, the Yellow and Yangtze and the Chinese, the Indus and the Harappans: all depended on fertilizing flood waters for agriculture. In each case, the society made the necessary accommodations to the threats posed by rising water. In some cases, dikes and canals were built; in others, patterns of settlement took into account the fact that any house might be flooded at any time during the rainy season. Furthermore, as sea levels rose in the post–ice age, the crops and buildings had to be protected not only from the water from snowy winters rushing down, but also from seawater pushed up by storm surges as the normal sea level inexorably rose. Undoubtedly, many lives were lost in years of particularly bad flooding, but on balance, the advantages outweighed the disadvantages, or the communities moved.

2

THOSE MYTHIC FLOODS

ndonesia Point says, "Jakarta has been endowed with only two seasons by ... Mother Nature, namely wet and dry. The rainy season starts from October and lasts till March. Dry seasons come from May to September."[1] However, according to an article appearing in the *Jakarta Post* on September 23, 2020,

> local authorities are preparing mitigation measures as they anticipate floods in the upcoming rainy season, after parts of Jakarta and West Java were already inundated on Monday.
>
> The Meteorology, Climatology and Geophysics Agency (BMKG) predicts that the rainy season — which often brings floods — will begin in late October or early November this year, but several regions have already reported torrential rain during the current transition from the dry to the wet season.
>
> The BMKG has warned of possible extreme weather during the transitional period.[2]

~

The story I mentioned in the last chapter — the one about the people who ventured across the Bab el-Mandeb Strait for the first time — was just that, a story. It happened so long ago that all traces have disappeared. But there are other ancient stories that can tell us a lot about our constant skirmishes with the sea.

Not related to sea-level stories, but pertinent nevertheless, is recent research by Australian anthropologists comparing stories around the world about the Seven Sisters, the stars of the Pleiades. In cultures as dissimilar as China in 2357 BCE, Germany in 1600 BCE, Greece in 700 BCE, and those in early Australia, stories about these stars revolve around a group of young women who are harassed by hunters or young gods. In all of the stories, there are seven, although it is the rare person who can see more than six stars in the group. In their article "Why Are There Seven Sisters?," Ray P. Norris and Barnaby R.M. Norris suggest that the stories all have their origins in a time, one hundred thousand years ago, when all seven stars were easily visible. Since then, the stars have moved in the firmament so that one of them is nearly obscured by another. They conclude their paper by saying that the story dates back to "before our ancestors left Africa and was carried by the people who left Africa to become First Australians, Europeans, and other nationalities."[3]

Not only do disparate cultures have stories that mention the Seven Sisters, a surprisingly large number of societies include a tale about a flood in their histories. Among the People of the Book (Jews, Christians, and Muslims, to take them in chronological order of their founding), the story of Noah immediately comes to mind: nearly all people are wicked, so the Lord sends a flood to wipe them out. There is one good soul, however, so the Lord warns him, telling him to build a big boat to save his family and useful animals.

The Greeks also had a story of a big flood. In it, Deucalion, son of Prometheus, the creator of humankind, and his wife weather a massive flood that Zeus brings down upon evil mankind. How old the story is isn't clear because it is part of the wealth of myths the Greeks told about their gods over centuries, if not millennia. The oldest surviving written version shows up in Book I of the Roman poet Ovid's *Metamorphoses*, in 8 CE, proof that stories were often around for a long time before they were written down.

Then there is the much older story of Gilgamesh, the Mesopotamian king who lived sometime between 2800 and 2500 BCE, or nearly five thousand years ago. His story — the *Epic of Gilgamesh* — is the oldest written literary text extant, and he appears to have been a real person who lived centuries before the story was recorded in writing: the name Gilgamesh shows up in other, older texts that give chronologies for kings in that part of the world. Therefore, the supposition is that the flood the story describes would also have occurred centuries before, with the story preserved by telling and retelling before it was pressed onto clay tablets. That means that the story dates to about the same time as the story of Noah and, perhaps, the first telling of the Greek stories about Deucalion.

The question arises, is it possible that these written stories were inspired by something that really happened? If you are a believer in the literal truth of the Bible, the answer has to be yes. That belief lies behind the many attempts since about 1850 to find the remnants of Noah's ark on Mount Ararat in the Armenian Highland at the extreme east of what is now Turkey. In 2010 researchers at a Christian think tank announced they'd found wooden structures that they identified as the remnants of the ark at a height of around 3,900 metres on the mountain. They said they had dated the remains to about 4,800 years ago, which would roughly coincide with the date implied for the Great Flood in the biblical text.[4] Publication of the research elicited scoffing from much of the scientific community for many reasons, yet many serious scientists are convinced that Noah's story and the other flood myths have bases in fact.

But you might ask, surely you don't think that people could more or less accurately pass along a story over generations? Just think of that kids' game Broken Telephone, where you sit in a circle and pass along a phrase by whispering it in your neighbour's ear. Usually by the time the run around the circle has been completed the message is garbled, sometimes to the point of being the opposite of what it was when it started.

Not so fast, though. Broken Telephone is just a game, and no one goes into it with a purpose more serious than having a good time. There are many societies in which oral tales are considered important, however, and because of this, stories have been passed along for generations. One case in point in Western culture is the whole suite of Greek literature. Homer — who wasn't one person, but who's going to quibble about that? — composed his epics in a time when the Greeks had no writing system. (That they didn't have one is a cautionary tale about what crisis can do to a civilization. The Greeks had developed a script now called Linear B, but about 1100 BCE, the society went through great turmoil and the system was lost.) It wasn't until sometime in the eighth century BCE that an alphabet based on the Phoenician one was developed and adopted.[5] The spoken word had pride of place, however, well into the time of Socrates (c. 470 BCE–399 BCE), who never wrote down any of his thoughts. Somewhat ironically, they were transcribed by his student Plato, and among them we find this comment on the effect of the development of a writing system: it "will create forgetfulness in the learners' souls, because they will not use their memories; they will trust to the external written characters and not remember of themselves."[6]

Nor were the Greeks the only ones to have a tradition that placed memorized stories and oral arguments on a plane higher than anything that might be written down. Indigenous cultures around the world have long kept their histories alive in the stories they pass from one generation to the next. And the oral tradition is prized in the widespread, long-enduring culture often called Celtic or Gallic. Take Julius Caesar's comment on the Gauls, made in 26 BCE in *The Gallic Wars*:

> They are said there to learn by heart a great number of verses; accordingly, some remain in the course of training twenty years. Nor do they regard it divinely lawful to commit these to writing, though in almost all other matters, in their public and private transactions, they use Greek letters. That practice they seem to have adopted for two reasons: because they neither desire their doctrines to be divulged among the mass of the people, nor those who learn, to

devote themselves the less to the efforts of memory, relying
on writing; since it generally occurs to most men, that, in
their dependence on writing, they relax their diligence in
learning thoroughly, and their employment of the memory.

Indeed, it was only in the seventh century CE that the great books of
laws in Ireland were written down, while *Lebor Gabála Érenn* (*The Book of
the Takings of Ireland*), better known as *The Book of Invasions* (and about
which more later), was finally recorded in about the eleventh century.

But the champions for handing down stories live in Australia. As we saw
in the last chapter, early modern humans arrived in Australia perhaps sixty-
five thousand years ago, and maybe even earlier, at a time when sea levels
were 100 to 150 metres below their current marks. Then, in a process that
was far from continuous, sea levels rose as the ice age ended. In the waters
around Australia, they reached their maximums between six thousand and
seven thousand years ago and until recently have remained relatively stable.
This means that any tradition or story about rising seas, if founded in memory, would have to relate to events that happened at least that long ago.

To investigate the possibility, a group of anthropologists has been examining both the geological record and two sorts of oral tradition handed down
by the Indigenous populations for eons. The first is a series of stories of what
the anthropologists call "narrated historical fact without obvious embellish-
ment."[7] The second are myths in which observed changes are attributed to
the acts of ancestral beings. These oral traditions were collected during the
first years of contact between Europeans and the Indigenous populations,
at the end of the eighteenth and the beginning of the nineteenth century.
They appear to be relatively accurate transcriptions of what the Indigenous
populations were saying in those early days of contact with Europeans.

A word here is necessary about the central place storytelling holds in
the culture of First Australians. Faithful recounting of geographic lore is a
key to their successful survival in such an arid land. Knowledge of watercourses, plants, and animals was and still is essential. It is bound up in
songs, which, when followed, can lead one through neighbouring country
and beyond.

The English writer and traveller Bruce Chatwin explored them in his book — part memoir, part anthropological text, part novel — *The Songlines*. Near the end, he notes that the main songlines in Australia appear to start in the north or the northwest, that is on the shores of the Timor Sea and the Torres Strait, which is the path that anthropologists now think people like us may have taken when they entered what is now a continent. Chatwin was writing well before recent research, yet nevertheless he observes,

> One has the impression that they represent the routes of the first Australians — and that they have come from *somewhere else*.
>
> How long ago? Fifty thousand years? Eighty or a hundred thousand years?

It looks as if he understood the journey that these first people made well before DNA testing and other new dating methods became commonplace.

One group of Australian First Nation stories recorded during the early days of contact with Europeans tells of how a single land mass became two, separated by water. Another group recounts how people could once cross a water gap, either by wading or swimming, but cannot do so today.

Among the examples of the first sort of tale is one about three islands off the Australian west coast that were once part of the mainland, with a lowland "thickly covered with trees." The story says that the trees "took fire in some unaccountable way, and burned with such intensity that the ground split asunder with a great noise, and the sea rushed in between, cutting off these islands from the mainland."[8] After determining how much lower sea levels would have to be for the water gap to be dry land, researchers have come to the conclusion that this tradition must date from at least 7,500 to 8,500 years ago.

Similarly, off the northern coast of Queensland, the edge of the Great Barrier Reef is about fifty kilometres from shore, but an early account states that "the Googanji natives ... say that before the flood the Barrier Reef was the original coastline, and that a river entered the sea near what is known as Fitzroy Island."[9] Yet for that island to be part of the mainland, sea level would have to have been twenty-three to twenty-five metres lower than it is today, and for the coast to be near today's barrier reef, sea level would have to have been about fifty to fifty-five metres lower. And when was the last time that happened? Well, sometime between 10,500 and 13,400 years ago.[10]

There are other stories, too, including one that does not concern rising sea levels, but which goes back even further, to the eruption of a volcano in southeast Australia some 37,000 years ago. To be sure, some scholars still doubt that oral traditions can be trusted, but these Australian researchers maintain that "there seems little doubt that [the stories] do actually recall the time when rising postglacial sea level attained the level of the continental shelf locally and rapidly inundated coastal lowlands."

They continue,

> Australian Aboriginal people's wealth of such deep time-depth stories stands out as exceptional among the corpus of such traditions globally, inviting speculation about what kind of cultural attributes might have fostered the accurate carrying of information in this way. The practice and nature of Aboriginal storytelling is likely to be key, particularly the ways in which it was ritually embedded in cultural practice, both routine and occasional, and the explicit teaching of "Law," the deliberate tracking of teaching responsibilities, coupled with high evaluation of unchangedness as a doctrinal principle, which encouraged the learning and onward transmission of particular traditions by successive generations.[11]

Okay, perhaps it is true that verses and stories can faithfully be passed down for generations, but is Australia as exceptional as these researchers maintain?

Three stories from widely separated parts of the world show that it is not. One example comes from China, where tales about a huge flood on the Yellow River are as much a part of that society's founding tradition as the story of St. Patrick's Christianizing mission is in Ireland. According to the stories, the Chinese flood occurred sometime about four thousand years ago. It lasted for months and only ended when Yu, a larger-than-life hero who became the founder of the Xia dynasty, tamed it by dredging. After being part of Chinese mythology for millennia, the veracity of the stories was questioned in the twentieth century, and for a time Yu's accomplishment was dismissed as mere folklore, not history. Yet in 2016 Chinese archeologists reported that they'd found evidence of the flood, including bones of children who must have been killed in it. Analysis of the debris allowed them to date the flood to about 1900 BCE, which reconciles historical and archeological chronologies.[12]

The example from the other side of the globe is not as generally well-known and recalls the Australian story about a volcano. In southern Oregon, the Klamath Tribes have stories of a rivalry between the spirits of Earth and Sky, who spent much of their time in two mountains, Mount Shasta (now the gorgeous cone-shaped mountain at the head of California's Central Valley) and Mount Mazama, about one hundred kilometres north of the Oregon-California border. Enter the beautiful daughter of the Klamath Chief, with whom the spirit of the Earth fell in love. She rejected him, and the spirit of the Sky agreed to protect her. An immense battle between the two mountains and their spirits ensued, with molten rocks being catapulted into the sky and rolling down the mountain sides. Finally, the spirit of the Sky forced the spirit of the Earth deep inside Mount Mazama, which then collapsed in a huge explosion that left a lake in the middle of the much lower mountain. It's now called Crater Lake and is held sacred by the Klamath People. And when was Crater Lake formed? Geologists have determined that the massive volcanic eruption that blew the top off Mount Mazama occurred about 7,500 years ago.[13]

There definitely were people around this area even before then: human coprolites — ancient, dried poop — covered with volcanic ash and found in caves not far away have been dated to before 12,000 BCE, a time when agriculture was just beginning and only a few people in the world had settled down in permanent settlements.[14] The folks who left the scat were likely the descendants of people who hopped from inlet to inlet along the Pacific Coast at a time when the great North American glaciers still covered much of the interior of the continent. There is evidence that semi-permanent settlements of fishers and hunters dating from 9000 BCE lie deep beneath the waves off the British Columbia coast. We'll return to this when we consider the history and probable future of the Salish Sea.

Then there's the ancient city of Dwarka in the Indian state of Gujurat. It is mentioned in one of Hinduism's oldest sacred texts, the Mahabharata, as being the home of Lord Krishna. The landscape is arid today, but offshore in the Bay of Cambay, research has turned up remnants of what was a much larger urban area, long submerged. While dating of the ruins is considered controversial, some artifacts appear to be seven thousand years old, and it certainly is older than the Mahabharata, which is thought to have been compiled about 400 BCE.

On the other side of India, the beauty of the temples at Mahabalipuram has been noted by visitors for centuries, leading to its sobriquet, the City of Seven Pagodas. The strange thing, though, is that there weren't seven temples, but two. Myths said that the others had been swallowed up by the sea, but that was generally discounted as nothing but a story, although

The temples at Mahabalipuram were legendary. Some were swallowed up by the sea.

fishers sometimes reported seeing what looked like structures on the bottom of the sea some distance from shore. Then in December 2004, the tsunami caused by that massive undersea earthquake off Sumatra swept across the Indian Ocean. It first caused the water to recede about five hundred metres on the Mahabalipuram shore, revealing the remains of a temple. When the huge wave crashed forward, the architectural remains were covered once again by water, but the turbulence also swept away masses of sand, which revealed what definitely was another temple. Research, including underwater reconnaissance, continues, but it's not known at what point the sea claimed the other buildings.[15]

~

All right, perhaps old stories passed from generation to generation reflect things that happened hundreds, even thousands, of years ago. But how does that relate to rising sea levels, and, for example, to the biblical and Assyrian floods? What do these tales say about how people reacted to these disasters?

The idea that rising sea levels during the great thaw at the end of the glacial period might be connected to these stories has attracted many researchers more scientific than those on the biblical fringe who have been looking for Noah's ark on Mount Ararat. The focus of their work is the Black Sea, which is connected to the Aegean Sea and the Mediterranean by a stretch of water called collectively the Turkish Straits.

Among the world's busiest in terms of trade and also one of the most dangerous passages in the world, it includes two relatively narrow channels, the Bosporus and the Hellespont. The Sea of Marmara lies between them. The passage is sometimes perilously rough. The effect of storms is complicated today by the fact that there actually are two water flows through the straits. One is a freshwater flow from the Black Sea to the Aegean, while at greater depth there is a much more saline current that flows in the opposite direction. At their narrowest, the straits are currently less than a mile wide and are considered to be the boundary between Europe and Asia.

The waterway has captured imaginations for thousands of years. The ancient city of Troy lies at its western edge; two legendary Greek lovers,

Hero and Leander, supposedly carried on a torrid romance across it; while the world-conquering forces of Alexander the Great and Xerxes, the Persian king, crossed it in opposite directions as they ventured into the realms of their enemy. More recently, swimming it became a challenge to adventurers like the English author Lord Byron, who wrote that Don Juan, the hero of his famous long poem, was such a good swimmer that "he could, perhaps, have passed the Hellespont / As once (a feat on which ourselves we prided) Leander / Mr. Ekenhead, and I did."

Byron's swim is commemorated every summer these days by his admirers — in addition to his literary fame, he became a champion and hero of the Greek Independence movement — but in the time we are talking about — some 9,300 years ago — the whole area of the Turkish Straits was high and dry. It was nothing more than a pass between what is now the Aegean Sea and the drainage basin of a lake that filled the bottom of the depression now occupied by the Black Sea.

For some time, it's been clear that rising sea levels boosted the water in the Aegean high enough to spill eastward through what are now the straits into the Black Sea basin. To study the sequence of events, researchers have used a number of methods including examining cores taken from the bed of the sea, the progression of organisms and plants living in either fresh or saline water, and evidence of rapid erosion on now-submerged land. The picture that emerges is of truly rapid change in the water flow. One study says that given the flow rates calculated for the time when the rise of sea level reached the high points of the Turkish Straits, the great gush of water flowing from the sea inland took place in "no longer than 40 years and possibly as little as a decade." This conclusion is corroborated by layers of detritus suggesting that the change came during "a rapid flooding event."[16]

After analyzing the organisms found in the cores taken from the Black Sea floor, the scientists involved in this study say this massive flood occurred at about 7300 BCE, at a time when the melting of ice caps resulted in floods and great rushes of meltwater in other parts of the world. Specifically, the melting of ice in North America resulted in a huge lake of meltwater dammed by ice that burst several times between sixteen thousand and eight thousand years ago, freeing enormous amounts of water to flow into the

Atlantic Ocean.[17] One of these massive pulses of meltwater found its way to the Mediterranean, breaching the sill that guarded the Turkish Straits and quite rapidly filling the Black Sea depression.

The initial breach of the entrance to the Turkish Straits was not the end of the story, though. Sea levels kept rising, including one period about 5,600 years ago (well after the initial influx of water into the Black Sea), when there is evidence that water levels rose by about a metre in a very few years.[18] If you've been paying attention to the dates, you'll notice that this is not that far off the dates for Gilgamesh and Noah.

How would people have reacted to the sudden rise of water levels? At the moment, we have no idea of the practical methods they might have used, although, as we shall see, there is evidence that people began fighting against rising waters long ago. But certainly, divine intervention would have been welcome, and, as a people looked back on their history and considered the fact that their ancestors were chased from their homes but had survived, they might have woven a tale about what happened. That is, we are good people and the proof of that is that divine forces saved us from a disastrous flood. The Bible goes a step further, saying that rainbows are the symbols of the Hebrew God's pledge that he would never again destroy the world by flood.[19]

~

All this occurred during a long period in which people were moving into territory in Europe, Asia, and North America that was newly liberated from ice, while others, farther south, were developing agriculture. There were new frontiers, green pastures, woods full of game — in short, a much bigger, more inviting world than had existed during the glacial peak. But we have evidence that some of that newly available territory welcomed people only for a relatively short time, geologically speaking. Indeed, at least one large region at the centre of what would become western Europe has disappeared and was literally off our radar until the mid-twentieth century. That's when archeologists became intrigued by items being dredged up in the North Sea by fishing boats and oil companies investigating likely places to drill for oil and gas.

Yes, the North Sea, that body of water on whose shores some of the most prosperous and populous countries of Europe front today. For decades fishing boats on what is called Dogger Bank (after the Dutch word for the kind of vessels the fishers used) had hauled up a wide variety of human-made artifacts, from harpoons to fish hooks to pots.[20] But it wasn't until the turn of the twenty-first century that scientists using such technical marvels as ground-penetrating radar and advanced computer imaging began to see what lies below the surface of the water. What they are finding is a country about a third of the size of present-day England, filled with lakes, forests, swamps, salt marshes, and mighty rivers that ran into a much smaller northern sea. Some of the artifacts indicate that people were flourishing there as far back as eighteen thousand years ago. As studies continue, it begins to look more and more likely that people — possibly tens of thousands of them at any one time — made the place home until rising sea levels covered the landscape completely about seven thousand years ago.

Doggerland doesn't sound all that inviting to us, today, but as archeologist George Nichols writes, "one reason twentieth-century [scholars] express surprise that prehistoric hunter-gatherers may have found wetlands attractive is that *they* do not."[21] Today, marshes and swamps mean mosquitoes; anyone who has walked, worked, or rested near one knows that little insects usually come with the territory, and their bites are a real irritation. Then there is what the mosquitoes might be carrying with them: Dengue? Yellow fever? Malaria? How could folks have thrived in such an environment, given the sort of diseases that you could catch in similar locations now?

But the people of Doggerland did not get malaria, because the disease didn't make it out of Africa until several thousand years later. After the disease jumped from animals to humans in central Africa between ten thousand and five thousand years ago, it only slowly moved north into Europe, because it was constrained by the range of the mosquitoes that carry it. (Some sources say that when it did arrive in Rome in the first century BCE, it was a turning point in human history. Disease borne by malaria-carrying mosquitoes that bred in the reclaimed marshes around Rome decimated the area in 79 BCE. The territory was not repopulated until the 1930s.)[22]

Even had malaria been present, though, it would have been only one of many dangers the people of Doggerland faced, but the rich hunting and gathering and fishing of the fens and marshes would probably have more than compensated for the deleterious effects of the disease. Marshlands are actually extremely productive, as we shall see when we talk about the various ways that North America was settled, parts of it literally one marsh at a time.

For a long time before the mid-twentieth century, there had been hints of this submerged Doggerland. Among them were tree stumps preserved below sea level in several places along the coast of Britain. Called Noah's Woods by some, they were long considered only a curiosity, although a book written about them before the First World War, *Submerged Forests*, presciently suggested that they were remnants of woodlands flooded by rising seas. Then there was the wealth of bones of wolves, hyenas, bison, horses, woolly rhinos, mammoths, beavers, walruses, and deer dredged up over decades by trawlers. All these finds suggested that below the cold North Sea waters lay territory that had once been full of plants and animals.

But the idea that they represented anything more than oddities was taken seriously by few.

Thinking began to change in 1931, when a harpoon-like object was found in about fifteen metres of water some forty kilometres from the nearest shore, suggesting that people with definite skills had lived on land now underneath the waves. Just how long ago they might have roamed there could have only been guessed. Coming up with more definite answers took two major advances in scientific thinking: analysis of the clouds of pollen rain that fell over millennia and were preserved in the soil, and a means of accurately dating just how old the pollen was.

If you look at layers of ancient pollen in a long vertical core sample of soil, you may get a clear picture of what plants were growing at what time, thus documenting changes in climate and — once people began domesticating plants — cultivation practices. Some of the most interesting research was done in the 1930s by British scientists who dug down into the soil in East Anglia.[23] The digs plunged more than five metres into the earth, finding evidence that the climate and vegetation changed more than once, from marshland with nearby ash and pine to a much drier landscape. Once scientists began looking at chunks of peat and other material — the fishers call it moorlog — brought up by fishing gear, they were able to find similar sequencing from under the waters of the North Sea. The evidence was such, as Grahame Clark wrote, that it was "eminently probable" that societies of hunters and gatherers "not only existed, but flourished" in the now-submerged landscape. "Moreover, there is good reason for believing that the coast of the old mainland between east Yorkshire [in Britain] and north Jutland [in Denmark] must have been especially favourable for settlement."[24]

However, for various reasons — not the least a world war — these interesting findings were pushed to the back of the archeological closet until the early 1960s. While it was becoming widely acknowledged that the North Sea bottom had once been dry land, most thought of it as a land bridge, a place by which the early populations passed from mainland Europe to the British Isles. The assumption was that places where there are now major centres of civilization were also more "civilized" back then, so few people tarried on the land bridge.

But as the finds continued to accumulate, the ethnocentrism of this view became increasingly apparent. Surveys of the North Sea bottom off the Danish coast have produced more than two thousand archeological sites and burials. At the beginning of the twenty-first century, it began to look as if the region was rich with cultures different from those of either mainland Europe or the British Isles. In addition, serious study of the seabed using radar and other surveying techniques has disclosed a vast river system encompassing the beds of today's Seine, Thames, Meuse, Scheldt, and Rhine Rivers; it ran southwest along what is now the English Channel.

One of the biggest sources of new finds corroborating the presence of humans in Doggerland is literally being unearthed as the seabed near the Netherlands is mined for material to make a defence against the rising seas. In addition, continuing research into the past of the Thames Estuary in Britain gives a glimpse of what life might have been like in Doggerland. This is a subject we'll return to later.

But note that there is practically no lingering echo of Doggerland in the stories people who live around the North Sea tell today. One might say simply, Well, that was a long time ago; what do you expect? But there are two very good reasons why all traces of this rather welcoming country no longer exist.

The first is because of a catastrophe of nearly unimaginable proportions. The second is because stories about a place are usually told by the people who live there, not by the people who left, and that has implications for our future.

First the disaster. As mentioned earlier when we were talking about seismic activity in the Ocean Isles, undersea earthquakes can bring about huge movements of water, causing waves that travel thousands of kilometres and wreak devastation wherever they hit land. Remember, the 2004 one killed about 225,000 around the Indian Ocean. But tsunamis can occur elsewhere, too. One of the worst that we know of began off the coast of what is now Norway when an earthquake triggered a massive underwater landslide along the edge of the continental shelf. Using a number of tools, researchers have documented a swath of destruction radiating west and south from the slide, which is called Storegga, the Norwegian word for "Great Edge." Layers of sand on the shores of Scotland and northern England bear witness to the load of sediment that the waves brought in and then dumped as they retreated. There is evidence that the waves raced across the North Atlantic, probably reaching the shores of Greenland. From the remnants of destruction, it appears that in places the waves were more than twenty-five metres high.

And when did this catastrophe occur? Around 6100 BCE, and, to judge from bits of twigs and other vegetable matter preserved in the layers of detritus, probably in the fall of the year.[25] From the available evidence, it appears that this was after the sea had been lapping higher and higher on Doggerland for several hundred years. Some studies suggest that by then much of the formerly hospitable countryside would have already been flooded and people must have been looking for new homes. The tsunami might have just been the last straw. As one group of researchers puts it, "It is conceivable ... that final abandonment of the remaining remnants of Doggerland as a place of permanent habitation ... was brought about by the Storegga tsunami."[26] It looks also as if the last connections between Europe and England sank between the waves at this time, with the British

Isles becoming distinct from Europe, in what some wags have called the first Brexit.

So, that's the disaster. What about the stories? Well, Norse mythology speaks of a great flood in the times of giants, but it is of blood, not of water. On the French shore, there are stories of sinking cities, as we'll see. But except for tales, which seem to be borrowed from the Bible traditions, stories of ancient mythic floods are very rare in the lands that would have been next to Doggerland. The one exception comes from Ireland, specifically from *Lebor Gabála Érenn*, which we mentioned before as an example of a memorized tale written down only in historic times.

In it, the first peopling of Ireland is attributed to a group led by Cessair, who was the daughter of Bith, who was a son of Noah. Their divine instructions were to head west to escape an impending flood, so they set out in three ships, two of which sank. The third reached Bantry Bay on the southwest corner of Ireland with fifty-three survivors — Cessair, forty-nine other women, and three men; all but one died when the flood finally arrived. The survivor does some shape-shifting, becoming a salmon, then an eagle and a hawk, but ends up living 5,500 years after the flood and in the end composing this history of Ireland.

A fetching tale, I suppose, but what I find striking in this context is the part about trying to escape the flood. Sure, Noah was trying to do that, too, but actually he doesn't go all that far. These Irish forebears were ready to go quite a long way to escape rising water — perhaps like the inhabitants of Doggerland. The tale ends sadly, though, since it takes six other "invasions" before Ireland becomes populated. It might be that this actually reflects the real story of how people came to Ireland; great migrations have always been fraught affairs. Getting away from disaster isn't easy, and blessed is the person who lives to tell the tale!

Flash forward to the present and then one hundred years or so into the future for a tale of what might happen when disaster becomes too much. The scene is a United States following a Second American Civil War, as imagined by Omar El Akkad in his prize-winning apocalyptic novel, *American War*. In it, Benjamin Chestnut, an archivist and historian, tells what happened during a time of plague and rising water from his vantage point in New

Anchorage, in what was once Alaska. His story is savage and brutal, involving a young woman who is a carrier of debilitating disease, and horrendous destruction due to war and climate change. But this telling would seem to be a dead end, because after he's written it, the narrator destroys the woman's notebooks as "the only way I had left of hurting her." One moral to draw from the tale is that if we're not careful, there may be no one around to tell our stories. Another is that stories are the cornerstones of civilization.

But it's essential to come to terms with rising waters, and that's what people have been doing for as long as we've been faced with them. As Jim Leary writes in *The Remembered Land: Surviving Sea-Level Rise After the Last Ice Age*, the people of Doggerland appear to have been able to switch from a dry-land to wet-land lifestyle. He insists that they should not be seen as victims, since they found ways to cope through diet change and migration. Marcel J.L.Th. Niekus and his colleagues agree in a poster from Rotterdam's Rikjsmueum van Odor. They write that the folks who made Doggerland home changed as the situation changed and frequently found that new conditions provided an opportunity rather than a threat.[27] Undoubtedly, the same thing occurred time and again as sea levels crept higher. The question, of course, is whether we can follow suit this time around.

PART 2

TAKING BACK THE LAND, LIVING WITH WATER

MUSICAL INTERLUDE

One of my favourite composers is Claude Debussy. I particularly like his Préludes, which are short works for the piano that are fragile and beautiful and simple enough for someone who isn't a big classical music buff to understand and appreciate. One of the most fetching is "*La cathédrale engloutie*" ("The Submerged Cathedral"). A little more than five minutes long, it opens with a section that, Debussy writes in the score, should be profoundly calm, as if the cathedral in question were slowly rising from the waves into fog. Bells ring out again and again, and the theme is played strongly, as if the piano were an organ. Then the cathedral sinks back into the water, with the organ growing muffled by the waves, and the bells chime distantly. A rendition played by pianist Henrik Kilhamn that gives you some idea of the piece's beauty and Debussy's aim can be viewed on Sonata Secrets, Kilhamn's YouTube channel.[1]

If you've stayed with me thus far, you may wonder, Just what is she on about? How does a short piano piece have anything to do with the threat to civilization from climate change and rising seas? But bear with me; it does.

3

FIGHTING BACK

I f there appear to be no stories about the societies that once prospered in Doggerland and then succumbed to rising waters, this does not mean that people there and elsewhere have not fought back. When it comes to Doggerland, we don't know what might turn up. In other places, we have evidence, both physical and in story, about how they did.

Let's take the physical evidence first. Off the shore of what is now Israel, the remains of a seawall built about 5000 to 5500 BCE as a barricade against rising seas have recently been identified at Tel Hreiz.[2] The ruins were first discovered in the 1960s but weren't studied intensively until a series of storms scoured away the sand that had covered the wall, which is at least one hundred metres long. The end is still covered by sand.

Made of boulders that must have been brought from at least two kilometres away, the wall lies on the seaward side of the village. The remains of many early villages now underwater have been documented in widely dispersed places, but this wall appears to be the oldest uncovered so far. Researchers say the village was probably originally built about ten metres above the sea, but that the rate of sea-level rise during the period was probably steady, averaging about four to seven millimetres a year. That may not sound like much but it all adds up: the change would be something like twelve to twenty-one centimetres over thirty years, definitely a difference that people would recognize — and worry about. It should be noted, too, that sea-level calculations are more or less averages; in times of unsettled weather and high tides, winds could have brought storm surges that would have run up the beaches much higher, wreaking havoc on the village.

Obviously, the seawall didn't stop the sea, but some First Peoples in Australia have stories that tell of successful efforts to halt rising water levels. As Patrick Nunn explains, the stories date from about the point when sea levels around Australia stopped rising for reasons that have nothing to do with human intervention, yet nevertheless, the happy coincidence has lived on in stories still told today. One set of tales comes from northeast Australia, where waters rose over the Great Barrier Reef, but were stopped, so it seemed, when people built fires along the shore and piled hot rocks along the water's edge. Another set of stories tell of rising waters on the other side of the continent. There it took the prompt action of birds weaving roots and branches together to halt the rise; in this story, people were helped by powers stronger than they were.[3]

Then there are the more practical interventions hinted at in other stories. The flooding of the ancient Indian city of Dwaraka has already been mentioned, but what wasn't commented on then was the fact that, obviously, people had already been building seawalls and barriers to keep out the rising sea, since the coup de grâce for the city came when the gates were opened.

Which brings us back to Debussy and the haunting music rising from the depths off the Breton shore.

The piece was inspired by the legendary Ys, a town on the coast of Brittany in France. It was protected from the sea by a series of dikes and seawalls whose gates were opened at low tide to allow water to drain from the land in a pattern we'll look at more closely a little later. But, so the story goes, the king's daughter was persuaded by a demon to open the gates at high tide. The town was flooded, all was lost, and now the bells of the drowned cathedral can be heard when the tide and the winds are right. Patrick Nunn, who has looked so carefully at oral traditions around the world, points out that the story suggests the town had already been under siege from the sea; it had by then built defences against the water. He also observes that there are numerous stories of lost cities off the Atlantic coast, in stark contrast to tales told about the North Sea coast, where the fight against the sea has been and is continuing to be valiantly waged by people whose ancestors might have been Doggerlanders.[4]

Take the story my grandmother told me when I was about six and so full of energy that today I might be suspected of having attention deficit disorder and being hyperactive. I couldn't sit still, and I went from one project to another, which my grandmother found annoying. I also wasn't very perseverant (six-year-olds frequently aren't) and that, she thought, boded not at all well. So she told me instructive stories, including one about a Dutch boy who saved his town from being flooded when the dike that protected it sprung a leak. Thinking quickly, he stuck his finger in the hole and stayed there all through the night, even though it was dark and cold. His arm was almost insensible in the morning when someone came by and, hearing his story, went away quickly for help. The intended moral, for me at least, was that a small action could save many, and that all children should stick at it even though they are uncomfortable and bored and afraid.

Did I learn the lesson? Not sure. But what I did learn later was that the story is not an old Dutch classic, but something "borrowed," to use a polite term, from a French writer. Mary Mapes Dodge published the story as part of her outrageously popular novel *Hans Brinker, or the Silver Skates*, in 1865. It was excerpted in McGuffey Readers, which were used over generations in American schools, and that is very likely where my grandmother heard the story. But the Dutch only heard about it when told by Americans.

I hadn't thought about it in years until I started looking at the Netherlands' long battle with the sea. Three things are notable about the story. The first is that the original story from which it appears Mary Mapes Dodge copied was written by a writer who spent quite a bit of time in the region of Bordeaux, where over a long period people have employed many schemes to keep back flood waters. Also interesting is the fact that a very similar story is found in Hindu holy books, suggesting that the importance of dikes and water barriers has been recognized worldwide for millennia.[5] Third, once you learn a bit about what it takes to hold back rising waters, you realize that a finger isn't going to be enough.

~

The Dutch effort to control the seas goes back more than a thousand years. At the moment, we have no idea what the country was like during the long period between the flooding of Doggerland and the point when written records and archeological finds begin to tell us about the North Sea coastlands at the time of the Roman Empire. Very likely, the underlying geology was much the same as it is today: some higher country in the south where melting glaciers left hillocks of sand and rock; farther north, periodically flooded peatlands through which some pretty big rivers run; a mixture of inland lakes and tidal inlets; and a sweep of sand dunes along the coast.[6] While there is evidence that people hunted and fished in the lowlands, archeological research suggests that the immediate coastal area was not home to many on a permanent basis.

When the Romans marched north beginning in the third century BCE, they found various tribes passing some, but not all, of their days there. It's telling that during the four centuries the Romans controlled the territory, their fortresses and settlements were built either on high ground or on carefully chosen places along what was then the main channel of the Rhine. Their biggest (and oldest) city, Nijmegen, is far from the sea, although smaller centres were closer to the coast, including at least one that has since completely disappeared due to rising sea levels. The salt marshes, fens, and extensive peatlands appear to have been basically unsettled. The Rhine system flooded periodically, and perhaps more importantly, the natural bulwark of sand dunes along the coast was occasionally and drastically breached during fierce winter storms.[7]

Great arcs of sand dunes have long swept across the southern shores of the North Sea, from the Pas-de-Calais, across northern France and Belgium, through the Netherlands, and into the northern part of Germany. They are the result of millennia of erosion by ice and water. Some of the sand has washed down from the faraway Alps via the Rhine and other rivers, while some of it is the remnants of rocks ground down by glaciers during the various glacial periods. Today, the ocean tides and currents, which tend to run from west to east along of the southern shore of the North Sea, spread the sand, aided by winds that can sometimes be ferocious. Researchers have found hints of what the shore looked like over the geologic history, but there are also written records of some of the movements of sand and sea over the last two thousand years. At times, these movements proved catastrophic for the people who had been sojourning on the land.

The first recorded instance occurred in the second or third century CE along the coast to the west of what is now the Netherlands. It appears that the dunes were breached near Dunkirk and the sea invaded the land lying behind them. The exact date isn't clear, but because Roman coins bearing the likeness of Quintillus, emperor in 270, have been found under a layer of peat, and in the absence of later objects, the supposition is that this "transgression," as it is sometimes called, occurred during a time when the Romans were present.

Archeologists in France and Belgium have discerned two other transgressions in the north of France and Belgium, interspersed with periods when the sea seemingly "regressed," or retreated, so that the formerly flooded lands stood above the high-water mark.[8] Why this occurred is not clear. The apparent withdrawal of the seawater could have been due to silt and sand deposited during floods that built up the land substantially over a period of a hundred years or so. Or, conversely, the shifting arcs of sand offshore might have moved so that protective dunes held back the water. Significantly, researchers who have studied the Netherlands intensively do not speak of transgressions or regressions; they have put together a broader picture of what has happened and is happening along the North Sea coast, where the threat from the sea and from river flooding is more or less constant.[9]

The work of Dutch archeologists provides a revealing look at more than four hundred years of Roman influence. Their finds, combined with written histories, help mightily in deciphering the past. At least ten historical texts by Roman commentators and historians from the time of Julius Caesar through the disintegration of the Roman Empire can be mined for clues. Early on, Roman observers noted a system of earthen bulwarks surrounding swampy peat fields in what is now northern Belgium and France. At the turn of the Common Era, other Romans built canals and dikes along the Rhine in an attempt to increase flow in the main channel so that their ships could penetrate as far inland as possible.

Some of the Roman texts refer to a tribe called the Batavi, who occupied an "island" in the middle of the lowlands through which the Rhine and its tributaries flowed. Julius Caesar writes of the "Island of Batavia," which appears to have been land surrounded by three of the branches of the Rhine, while Tacitus, another Roman historian, calls the Batavians a formidable clan, whose men were prized conscripts for the Roman forces. They allied themselves with the Romans and "fared better than most tribes allied to the stronger power."[10] In *The Histories*, he writes, "After a long training in the German wars, they still further increased their reputation in Britain, where their troops had been sent, commanded according to an ancient custom by some of the noblest chiefs. There still remained behind in their own country a picked troop of horsemen with a peculiar knack of swimming, which enabled them to make a practice of crossing the Rhine with unbroken ranks without losing control of their horses or their weapons."

But that rather positive situation didn't last. The Roman leader Gaius Julius Civilis organized a banquet for the chief nobles and "the most determined" tribesmen. Tacitus writes,

> When he saw them excited by their revelry and the late hour of the night, he began to speak of the glorious past of the Batavi and to enumerate the wrongs they had suffered, the injustice and extortion and all the evils of their slavery. "We are no longer treated," he said, "as we used to be, like allies, but like menials and slaves.... Now conscription is upon us: children are to be torn from parents, brother from brother, never, probably, to meet again. And yet the fortunes of Rome were never more depressed. Their cantonments contain nothing but loot and a lot of old men. Lift up your eyes and look at them. There is nothing to fear from legions that only exist on paper. And we are strong. We have infantry and cavalry: the Germans are our

kinsmen: the Gauls share our ambition. Even the Romans
will be grateful if we go to war."

There followed an attack, which left the Roman forces decimated and
their forts in ruins. Civilis urged on neighbouring tribes: "Heaven helps the
brave. Come, then, fall upon [the Romans] while your hands are free and
theirs are tied, while you are fresh, and they are weary."

The conflict pitted other Gallic and Germanic tribes against the Romans,
who eventually prevailed. But what remained in the Dutch national mythology was the memory of strong resistance to outside forces. In the sixteenth
and seventeenth centuries, when the Dutch fought for their independence
from Spain, they came to identify with the ancient Batavians. During that
long war, politicians, writers, and artists used the example of the Batavian
revolt to encourage the Dutch cause and its fighters. In the late eighteenth
century, the Dutch even called the republic they established the Batavian
Republic, but before then, they used the name Batavia for the outpost of the
Dutch East India Company that is now Jakarta.

By then, the Dutch had begun to come to terms with their precarious
position on the North Sea. Some of the research undertaken by Dutch and
other scholars suggests that for a few hundred years following the decline of
Rome, the whole North Sea coast became much less populated. Then, like
a wave rolling back toward the shore, some people returned as confusion
and conflict increased in what was the former Roman Empire. By the early
Middle Ages pressure on easily farmed land was increasing, which led some
canny farmers to turn to the vast fertile peatlands and marshes formed over
centuries of retreating and advancing floods from rivers and attacks from
the sea.

Efforts to bring the peatlands under cultivation began in earnest about
the eleventh century. At the time, much of them were above sea level, which
itself was a dozen centimetres or more lower than it is now. Therefore, the
work was rather straightforward. Canals were dug, through which fresh
water could drain from the fields into streams and rivers, and thence to the
sea. Things became more complicated as time passed, however, because of
the way peat ages and the way people put it to work.

Windmills provided the power to pump water from fields.

Peat — plant material that has decayed over hundreds if not thousands of years — is sometimes described as "young coal" because it is made of plant matter like coal and might, in fact, become coal in a couple of million years under the right circumstances. Before then, though, it can be burned, and for hundreds of years leading up to the Industrial Revolution and the switch to burning coal itself, peat provided an easily accessible fuel. While that practice kept Dutch (and other) peasants warm, it also meant that peat from drained fields was frequently dug up to burn, which lowered the level of the land. Worse, once the peaty soil is exposed to air, it begins to change in chemical composition and to compact.

In the Netherlands, what was grown on the peat was taken off the land, and the soil was not replenished with new sediment and organic matter. The result was that within a couple of hundred years, the transformed peatland was no longer higher than the rivers. That meant that new ways to pump water off the land and into the canals had to be found. Windmills, now such a symbol of the Netherlands, were developed to do just that. The water the windmills pumped into the canals — whose sides were by then being reinforced and built higher because of the difference between the water level and the land — went first into reservoirs and then, at low tide, into the rivers, where it was discharged into the sea. The areas of reclaimed land, called polders, have served as examples for similar efforts around the world to save and/or reclaim land, as witnessed by the number of languages that have borrowed *polder* to refer to a piece of low-lying land reclaimed from the sea or a river and protected by dikes. Even the French use the term, although there's evidence that they have been making polders as long as or longer than the Dutch, although far less intensively.

The lowering of the land intensified the problem of flooding from the sea. This led people to build dikes specifically to keep out tidal water. These originally were local efforts — sometimes even built by individuals — but the complexity of the task mandated the establishment of water boards charged with building and maintaining defences against floods of whatever sort, supposedly weighing "all interests and regulat[ing] the water table accordingly." Since their first appearance in the fourteenth century, the water boards have become the cornerstone of all the considerable work that has led to the Netherlands' resistance to the sea. This cooperative approach is also part of the kit of techniques that the Dutch are currently marketing around the world as sea levels rise everywhere.[11]

And the North Sea is a formidable opponent. Various historians and researchers have drawn up lists of the big storms, which appear to have arrived at intervals of no more than twenty or thirty years, and some of which were responsible for thousands of deaths.[12] One of the best documented took place in November 1421. The storm was enormous and also wreaked havoc in England, where it completely reconfigured parts of the coast. Dunwich, once a major port, was largely washed away, and the town's cathedral was submerged; its bells, like those of Ys, are said to be still heard today. In the Netherlands, the powerful winds sent seawater crashing though dikes and up rivers to flood what is now southwest Holland. Polders were flooded and twenty-five villages disappeared under the water. So affecting was the flood that two generations later, parishioners of Grote Kerk in Dordrecht (a town mostly spared damage) commissioned an altarpiece commemorating the disaster. It now is displayed in the Rijksmuseum in Amsterdam.

Altarpiece commemorating the St. Elizabeth's Day flood in 1421.

But that was hardly the worst of the storms that surged into the low country. That probably occurred on December 14, 1287, the day following the feast of Saint Lucia, after which the flood is named, Sint-Luciavloed. It's estimated that between fifty thousand and eighty thousand people died when the storm surge roared over the sheltering dunes in what is now the northwest Netherlands and along the North Sea coast of Germany. There had been a large freshwater lake near where the river Vlie entered the sea, but the storm and the extraordinarily high tide swept away the dune barriers. Literally overnight, the lake became a bay, creating what came to be known as the Zuiderzee. The town of Harlingen, which had been landlocked, became a coastal town, while other villages completely disappeared. The flood would have long-lasting effects on the Dutch landscape, as well as on the Dutch people's attitude toward how to defend the Netherlands from the sea.

This great change in the coastal geography also opened the way for Amsterdam to begin its ascent as one of the most important cites in northern Europe. The swampy peat fields along the Amstel — then just a stream flowing into the larger river IJ — began to be reclaimed at least as early as 1085, records show. Then storms over a couple of centuries set the stage for the river to become something more impressive. A massive storm on All Saints' Day in 1170 breached dikes and sand dunes and transformed the hollow where the Amstel ran into a broad estuary. Upstream, the Amstel flowed more quickly, making it easier to drain the peat fields. Then came the flood of 1287, which resulted in Amsterdam's becoming a sheltered and accessible harbour with a straight shot to the North Sea. That was not enough for the Dutch, though; in the 1860s, a shipping canal was built between the city and the sea, across the narrowest part of North Holland, making access to the North Sea even more direct. It was another example of the way the Dutch were taking control of their environment.

Nevertheless, a word should be said here about climate change in the context of the Netherlands' later colonial adventures, including the Dutch East India Company's activities in the Ocean Isles. What is now the Dutch homeland has never been immune from wider climate tendencies. The period of the sixth century is a good example. It was one of difficult weather caused, it seems, by two big volcanic eruptions. The first one occurred in late 535 or early 536 and spewed so much matter into the atmosphere in the Northern Hemisphere that the sun was dimmed for a year and a half or so. Then, only about four years later, another huge eruption sent more matter into the atmosphere than did the one in 1815, when Mount Tambora in the Ocean Isles blew its top, creating a worldwide "year without a summer." The result was shortened growing seasons and erratic weather that caused widespread crop failures. While little evidence tells of the extent of the damage in the Netherlands, researchers are now pointing to the eruptions in the sixth century as aggravating societal unrest all over the Northern Hemisphere.[13]

Collateral damage appears to have been caused by the first bubonic plague — called the Plague of Justinian, after the Byzantine emperor who headed the eastern section of the Roman Empire following the fall of Rome. As volcanologist Clive Oppenheimer writes in *Eruptions That Shook the World*, "The argument goes that poor harvests drove the plague's vector, rats, out of the fields and into grain stores where they came into proximity with people." The disease raged around the Mediterranean, in some places killing as much as 20 percent of the population.

Stop there for a moment. When I read that, I was whipped back to what we were living through as I researched this. In Montreal, where I live, the death rate from Covid-19 has been nowhere near that high, even though in the early months of the pandemic, death rates soared as the disease ravaged long-term care facilities. One of the few good things about these times of Covid-19 has been the glimpse that it gives us of what life must have been like back then. We can see, if only dimly, what people did — and what they did was to continue.

The effects of the sixth-century volcanic eruptions did not touch only Europe. They appear to have caused crippling winters and other weather disasters elsewhere in the world, including events that led to the fall of the Mayan Empire in Central America.[14] But then, rather wonderfully, the weather got better, although the whys of this are only now being ferreted out. Suffice to say that beginning about 800 CE and continuing well into the thirteenth century, temperatures in northern latitudes rose until grapes could be grown in England and wheat on the Scandinavian peninsula, at 64 degrees north latitude, only a couple of degrees south of the Arctic Circle. There are indications that the prosperity contributed to the wave of cathedral building in Europe. An example is Chartres, near Paris, which was begun in 1145 and welcomed up to ten thousand worshippers on feast days.

It didn't last, though. Temperatures appear to have begun to fall at the very beginning of the fourteenth century. Possibly this was due to a reverse greenhouse gas effect that illustrates how human actions are far-reaching: following the Black Death, the Mongol invasions of southwest Asia, and the decimation of the Indigenous populations in the Western Hemisphere, forests encroached on previously cultivated land, sopping up CO_2.[15] Because of the cooler temperatures, no more wheat was grown at high latitudes — Brian Fagan writes that the diet of most folk in northern Europe returned to soup and gruel and porridge, "as they had done in prehistoric times."[16] The ambitious explorations of the Vikings, begun in earnest in the ninth century, came to naught; the western portion of the Greenland colony was partly abandoned by 1350, apparently because of bad weather, and by 1500 the larger eastern settlement was empty.

The colder weather of this Little Ice Age also had far-reaching ramifications when it came to the relation between people and the sea. Storms raged, and the Zuiderzee was not always a safe haven.[17] The idea of building a barrier across its mouth was first proposed as early as 1667, and the idea was raised several times afterward. More recently, the floods of 1916 led to Zuiderzeewerken, an immense project that cut off the Zuiderzee from the North Sea, making it a freshwater lake and producing a barrier against future storm damage.

The ambitious Dutch system of polders, canals, and holding ponds has also at times been used as a defence against human enemies. One of the most celebrated instances occurred in the late sixteenth century, at a time when the Dutch were in open rebellion against King Philip II of Spain, who had inherited the seventeen provinces of the Netherlands as a result of an arcane and complicated succession. The Dutch were not pleased by this, particularly because in a time of religious unrest many of them had become Protestants, while Philip II was militantly Catholic. Fighting roared on for years — the conflict eventually came to be called the Eighty Years' War, since it lasted from 1568 to 1648 — and in 1573, the Spanish began a siege against Leiden. The town, which had been besieged earlier, had not reprovisioned, and almost gave up after months of being surrounded by the enemy. But William, Prince of Orange, ordered the dikes opened so that the sea could flood the low-lying country. On October 3 his small flotilla of flat-bottomed boats was able to surprise the Spanish and take the town. The date has become one of the city's most famous holidays.

I have never been there to take part in the festivities, but I did get to visit another result of that famous victory. The people of Leiden, who had so successfully and stoically endured the siege, were offered a choice as a reward — a tax holiday for three years or the establishment of a university — or so the story goes.[18] Many would say they chose wisely because they chose the university. Established in 1575, Leiden University and its botanical garden, the Hortus botanicus Leiden, are among the oldest in Europe. Today, it is a green island in the city, which is linked to the rest of the Netherlands by roads and trains and — of course — canals. When I toured the garden in 2000, I was blown away by the wonderful re-creation of the garden where Linnaeus did some of his most intensive work in setting up the taxonomy system for plants, the system that underlies the one we use today. Had I known where my wanderings would lead me, I would have paid more attention to the ways in which this town could have been opened up to rising seas in an effort to save it by destroying it.

One other thing I remember clearly from that trip to the Netherlands was the lovely afternoon we spent walking around The Hague, the country's third-largest city and the home of the Dutch parliament. Trolleys, buses, and bikes crowded many of the streets, but canals clearly dating back four hundred years still criss-crossed the city. The apartment where we stayed was steps away from the tramway, and the bike path ran just at the corner, but out in front lay the canal that must have, not too long ago, been used to travel through this neighbourhood, which was built up shortly before the First World War.

Like other cities whose basic layouts date from before the automobile, The Hague is very walkable, and one day's amble included a short hike up the trail to the dunes that protect the city. Recently, when I talked to my companions about that day, one said that obviously the dune wasn't artificial because it was so big, so solid. That's true, up to a point. Dunes are such a natural feature of the coast that many place names from the eastern Netherlands to Calais feature them: Dunkirk, which gave its name to a series of sea transgressions and from which British forces retreated after the Germans invaded Belgium and France, comes in fact from a church — kerke, in Flemish — built on the dunes.

The beach we found over the top of the dunes extended for miles and miles. It was great for walking, and, we were told, wonderful for wading and swimming when the weather was warmer. But despite its solidity, the

Dutch were even then making plans to reinforce this dune defence. I have since learned that doing so is part of the current version of the Netherlands' long-term plan to protect the country from rising sea levels.

Dunes and sandbanks shift with time and tides, and part of the Netherlands' strategy for fighting back the sea relies on replenishing them. Chief among the twenty-first-century tools for doing this is the Zandmotor, a huge dredging project that displaced 21.5 million cubic metres of sand to create an artificial peninsula near The Hague. The idea was that the prevailing winds and current would spread the sand out, so over about ten years it would fill in the space between the peninsula and the shore, reinforcing the dunes and eliminating the need to dredge every five years. The massive project was completed in 2011, and so far, it seems to be working as predicted. It also has created another space for Netherlanders — residents of one of the world's most densely populated countries — to enjoy a little open space. One of the unexpected collateral benefits of the project has been the dredging up of artifacts from Doggerland, among them an ax that appears to be the handiwork of a Neanderthal and dated to some fifty thousand years ago.

So, the Dutch continue to put much effort into planning for a future with rising water. To understand what this means both here and elsewhere, it is good to look at the methods used to protect the Netherlands from what weather wonks call a five-hundred-year storm or a one-thousand-year storm.

The two terms are relatively recent additions to our weather talk. Since the 1960s, civil engineers, city planners, and insurance companies in North America have used the idea of a one-hundred-year flood both to weigh the risk of flood damage in particular places and to plan how to deal with raging water. The analysis is based on water flows, topography, and, to some extent, history. But labelling an event as a one-hundred-year flood is not the correct way of referring to the concept. Rather, the calculation is based on the idea that a certain outcome has a 1 percent probability of happening each year; in actual fact, a flood of this level might occur at any time, at ten- or fifteen-year intervals or at much longer ones, depending on the way the weather gods operate.

The U.S. Geological Survey periodically modifies its categories, updating with new data and weather models. These include the possible effects of more runoff due to increased construction, because roads and hard surfaces keep water from percolating into the ground and send more of it into streams and rivers. In the future, changing weather patterns are also going to have to be considered. But one of the take-aways from a look at weather history is that the probabilities of storms are not based on hundreds and hundreds of years of experience. In North America, people have been keeping detailed written records of floods and rainfall for only 150 years or so.

Dunes are what might be called soft defences against the sea. Those earlier transgressions during severe storms demonstrated the dunes' importance across the southern shore of the North Sea. A massive storm in 1953 showed, if anyone ever doubted, that much more is needed to protect low-lying countries when the sea is on a real rampage. Descriptions of what happened on the night between Saturday, January 31, and Sunday, February 1, 1953, are truly catastrophic. They show that once dune barriers are breached, people could well be drowned in their beds.

Certainly, the 1953 storm was literally not on the radar. Remember, this was less than eight years after the end of the Second World War, when the science of weather forecasting was considerably less developed than it is today. Predictions were focused on barometer pressure, cloud observations, and precipitation levels, and a curious thing noted during the war had not yet been put into action.

Radar, born to track movements of airplanes, works by sending out pulses of radio waves that bounce back from objects in the skies. It worked well when tracking aircraft, the war effort found, but sometimes the results were obscured by "clouds," which literally turned out to be clouds of falling water droplets — in other words, rain. The scientists seconded to the war effort noted this, but because of the press of battle, they couldn't investigate the phenomenon further. After the war, however, ways were found to turn a curiosity into a valuable tool.[19] Today, we can casually call up the radar maps maintained by most weather services on our cellphones. When evacuation orders come, they are based on solid evidence of approaching danger, even though weather can still throw a surprise or two our way.

Those pictures of TV weather people doing stand-ups in front of pounding surf and palm trees bent nearly double by hurricane winds are commonplace now, but it was only in 1961 that the first radar image of a hurricane bearing down on land was broadcast. The enterprising TV reporter Dan Rather, then still in his twenties, convinced his bosses and the local weather office in Galveston, Texas, that superposing an image of the Texas shore on the radar screen showing the progress of Hurricane Carla would be both effective weather service and good television. The result was that some 350,000 people took an evacuation order seriously and headed away from the coast. Only forty-six lives were lost, even though a smaller storm that made landfall in nearly the same place sixty years before resulted in something between six thousand and twelve thousand deaths.[20] Since the 1960s, weather expertise has increased: radar is now coupled with satellite imagery to give an even earlier and better look at developing storms.

In the winter of 1953, however, there was nothing of the sort, and both the severity of the storm in the North Sea that night and its path were unknown.

The windstorm, accompanied by a very low-pressure area and an unusually high tide, created a storm surge that in places was 5.6 metres above mean sea level. More than 1,800 people were killed in the Netherlands, and more died in Belgium and the United Kingdom, while it was estimated the thirty thousand animals drowned. In all, 9 percent of Dutch farmland was flooded by seawater.

One evening when I was working on this chapter, I spent some time looking at the film taken in the aftermath of the disaster. I could find none from the night thereof, however. The reasons for this are numerous, not the least because the worst came in the night, and even when warned, it's unlikely that photographers would have been able to capture the menace, given the technology of the day.

Weather warnings were sent out late Saturday, but because radio stations in many areas did not broadcast after sundown, that method of relaying the message didn't function. Telephone warnings weren't effective either, given that many households, particularly in rural areas, weren't hooked up. What was left was the wisdom of the crowd, of people who had seen big storms before and had some idea of how to react.

The damage could have been worse had it not been for a desperate move on the part of the mayor of Nieuwerkerk, who commandeered a riverboat and ordered the owner to plug a hole in a dike that had begun to slump under the pounding of waves and wind. The skipper battled the wind and rising water, and, luckily for the three million people who lived behind the dike, successfully rammed it sideways into place.

Clearly a finger in the dike would not have worked this time. What was required was a dangerous manoeuvre on the part of people who were accustomed to contending with the sea. Farther to the east, the dikes across the entrance to the Zuiderzee held, which, many say, more than paid for the cost of their installation at the beginning of the twentieth century.

When I'd finished shivering in sympathy with the cold folks fleeing their submerged villages, I went looking for more recent footage, particularly of flooding in Jakarta.

And there I found a plethora of sad sights.

~

February 22, 2021

Floods cause widespread disruption in Jakarta

Heavy rains in the capital city and surrounding areas from Friday to Saturday triggered floods that have deluged large parts of Greater Jakarta and disrupted the transportation system across the capital.

The flooding forced thousands of Jakartans to take to shelters and it is not expected to recede soon as the weather agency has forecast that the extreme weather conditions are expected to continue for the next several days.

The Jakarta Disaster Mitigation Agency recorded nearly 200 neighborhood units (RT) across the capital inundated by floodwater as of Saturday morning with 1,361 people forced to take shelter away from their homes.

The flooding ranged from around 0.4 to 2.5 meters in depth, with East and South Jakarta the hardest-hit areas.[21]

~

January and February are supposed to be the height of the rainy season in Jakarta. That is why, even though I had my travel grant money in hand by January 2020, I didn't plan a trip to Indonesia until April or May. Downpours were to be expected; floods were likely. And nobody wants to spend a limited research trip taking shelter from torrential rains or cowering under an umbrella.

Not going then was a good call, as the floods that January were legendary. As we saw earlier, in 2020 the rainy season picked up again a couple of months before it usually does, so there were floods in September also. Then came the rains of February 2021. I spent several hours watching videos and reports of the flooding, in which at least five people died. When I finished, I wondered that there were so few fatalities.

There were several differences between these floods and the disastrous ones in the Netherlands that prompted development of that country's

Jakarta has been prey to floods forever, like this one in 2016.

effective defences against water. First, there was the difference in climate: unlike the Dutch shivering in the winter cold, the Jakartans were dressed in much lighter clothing, since the high temperature on the day of the biggest flood in early 2021 was 28°C, with a low of 24°C.[22]

A bigger difference, however, was the way that people in Jakarta seemed almost resigned to the inundation. The Dutch films showed people trudging along the high ground between polders, dragging their possessions behind them, leading farm animals, looking almost shell-shocked. Subsequent history has shown that they were able to channel their sorrow and concern into rapid, effective projects that so far have stopped destruction. Films of the victims of Jakarta's floods, on the other hand, included shots of people who were up to their waists in water, but were acting as if they knew just what they should be doing to save themselves and their possessions. It's significant that four of the five people killed in the February 2021 Jakarta floods were children who had been swept away and drowned, not adults who had a good idea of how to save themselves.

This kind of flooding has become almost commonplace for Jakartans in recent decades, which is part of the reason why moving the country's administrative capital to another island looked so good to the government. The forty-billion-dollar price tag the project initially was given may actually be less than it would cost to come up with an integrated plan to keep Jakarta from being drowned on a regular basis. As it stands now, the millions of people who will be left behind when the capital moves were left out of the equation.

Jakarta is faced with three kinds of floods. (It is not unique in this; as we'll see, other cities and regions are also hit by all three, but as in many other things, what happens in Jakarta illustrates well what happens elsewhere.) The first occurs when great quantities of rain fall upstream from the city, overflowing the lakes and marshes where rainwater usually collects. The result is walls of water that fill the rivers running into the Java Sea. The second sort of flood follows heavy local rain falling in the city itself, which overwhelms the capacity of canals, drains, and rivers to carry water to the sea. The third is flooding from the sea, when dikes along the waterfront collapse or are overtopped by storm surges.

In January 2007, a flood of the first type devasted Jakarta, producing more damage than any flood in a century. Then, in November of that year, sea dikes failed and prompted the government of the day to start the process of rethinking the way that the shoreline is protected. The result was the National Capital Integrated Coastal Development (NCICD) plan, elaborated in conjunction with the Dutch (of course!) consulting and construction firm Deltares. The idea was to build a giant seawall about thirty-two kilometres wide, where an airport, toll road, residential area, industrial park, and waste treatment plant would be installed, as well as harbour facilities and green areas.

Taking an idea from Dubai, where reclaimed land forms artificial islands that look like a palm tree from the air, the original plan was to make Jakarta's great wall look like a mythical bird, the Garuda, which has become an emblem of Indonesia. In all, reinforced and new coastal wall defences would run about 120 kilometres along the northern shore of greater Jakarta. The idea is that not only would the force of rising water and storm surges be reduced by the works, but also, eventually, the lagoons behind the seawall could be cleansed and transformed into freshwater reservoirs that would provide clean water to the city, where less than half of the population had good, piped water in 2020. Work on the NCICD began in 2014 and continues, with the government pledging to persevere despite the Covid-19 pandemic, which has strained the country's finances. Unfortunately, the project has come up against some large problems, including the collapse of a 167-metre stretch of the previously existing seawall in December 2019.[23]

The second type of flood was supposed to be prevented by an elaborate effort to clear and widen the watercourses in the city and its surroundings, but obviously the projects have not been successful, as witness the floods of February 2021. Following those floods, city officials and their critics blamed each other. The widening and concreting of river channels was not working as a flood control measure, some said, while others said an alternate plan of "naturalizing" the watercourses would take too long and would be ineffective anyway.[24]

Significantly, relatively few people were talking about a fourth factor contributing to flooding in Jakarta: the subsidence of the city due to groundwater being sucked from under it to meet the needs of the megacity.

What this means in day-to-day life is brilliantly shown in a documentary by the Singapore public broadcaster CNA.[25] In it, Nemin — "I don't know how old I am, probably about sixty" — and his son Ujang set out early one morning from their home in a village-like suburb of Jakarta to drill a well in a similar neighbourhood. The men carry all their equipment, including metres and metres of coiled piping, on the back of a motorbike. They have an electric generator with them to power a pump, but muscle power does the drilling itself. If they reach water, they'll earn the equivalent of about thirty-five U.S. dollars. If they don't, they won't get paid. They offer no guarantee on the quality of the water, either. That is probably a good thing from their point of view, since it's going to be pretty awful. Given a penury of sewer treatment, the groundwater is almost certainly contaminated. Yet, as the narrator comments, going this route is a more cost-effective way for thousands and thousands of Jakartans to get water than to buy it either in bottles or from water trucks.

It takes the two men about ten hours of effort before they finally strike water — a delight for all concerned. What isn't good news, though, is what this kind of wildcat drilling is doing to the water table. Ujang points out that big apartment blocks and buildings where middle-class people live and work pump a lot more water out of the aquifer than do wells like the ones he drills. Less than 40 percent — mainly in the wealthier central sections — can count on treated and piped water. Until that service is expanded, people are going to continue to pirate water and contribute to the sinking of the city.

According to Intan Suci Nurhati, a researcher at the Indonesian Institute of Sciences, groundwater depletion means that the ground subsides up to eighteen centimetres a year. Even plans to beef up coastal dike defences may be endangered by this, since sinking may literally undermine bigger, heavier seawalls and dikes. The weight of buildings and infrastructure contributes to subsidence as the soil is compacted, but the real culprit is water mining. The documentary ends with photos of concrete walkways and bridges that have cracked as the ground has shifted.

Unfortunately, to date few techniques have been developed to reverse ground subsidence due to groundwater depletion. Nurhati says one of the things that's needed is to increase the ability of rainwater to penetrate the ground. Ujang also notes that his neighbourhood is much more paved over than it once was, with far fewer trees, fewer ponds, and fewer places for rain to percolate into the soil. Some experts say that setting up polders to store runoff may help, while another plan for Jakarta calls for the construction of 1.8 million vertical drainage units — sort of reverse wells where water can be pumped into the layers of subsoil, which can store water in their pores. The efficacity of that is open to question, and at the moment, only about 200,000 have been installed anyway.

Far more successful is what Tokyo has done over the last sixty years. Faced with rapidly depleting groundwater resources, the Japanese metropolis effectively shut down excessive water withdrawals by rethinking water usage and tapping into water resources outside the metropolitan area. A major part of the program was developing opportunities for rainwater to percolate into the aquifer. We'll talk much more about this later, but for the moment, just note that few efforts like this have been undertaken in Jakarta to date. The current government pledges that the entire Jakarta region will be served by good-quality tap water by 2030, thus eliminating the need for private water wells, but don't hold your breath. The take-away is that until this is done, there will be no subsidence in the subsidence, and the flooding crisis will continue.

It's possible that this uncertainty is behind investments, by foreign concerns, in other ports on Java besides Jakarta. In March 2021, I woke up one morning to find a news story reporting that the Caisse de dépôt et placements du Québec (CDPQ), the pension fund of the province where I live, was going to invest $1.2 billion, along with an Indonesian industrial giant and a Dubai-based shipping firm, in port facilities at Gresik, on the other side of Java. It didn't get much play in media around here, even though the CDPQ has managed Quebec's public pension plan since the 1960s, which is worth Can$365.5 billion. In effect, it was just another investment for the fund on which I and several million other Quebeckers depend for our retirement. But Gresik, Indonesia? Why that port? Why not something in Jakarta?

I wasn't able to get more than the usual press release blah-blah from the parties involved: "Through this partnership with Maspion, CDPQ is delighted to make its first infrastructure investment in Indonesia, a strong growth market which benefits from favourable structural trends," said Emmanuel Jaclot, executive vice-president and head of infrastructure at CDPQ.[26] No comment, in other words, that might shed some light on the future of Jakarta.

But looking further, it's clear that what is going on is the kind of thing that has been current since people started shipping goods around the world. Gresik — now a satellite town to Surabaya, Java's second-largest city — was once a major port. The Portuguese adventurer Tomé Pires, who travelled in these waters in the early 1500s, called it "the great trading port, the best in all Java.... This is the jewel of Java in trading ports. This is the royal port where the ships at anchor are safe from winds, with their bowsprits touching the houses. It is called the merchants' port; among the Javanese it is called the rich people's port."[27]

Located near where Java's longest river, the Bengawan Solo, meets Madura Strait, Gresik was the place where rice grown on the rich soil of the long river valley was traded for spices and other coveted items found on the islands to the east. It became home to a large colony of Chinese merchants and traders and was a centre for Islam in Java from the eleventh century onward. But it fell from prominence as the power shifted among the warring kingdoms of the interior of Java. In addition, the load of sediment carried by the river and its branches also contributed to slowly making Gresik much less attractive as a port. The silt made the river shallower and shallower and rendered more difficult the passage of the small boats transporting crops. Because of this, in 1890 when Surabaya — located where another river enters the strait — became a major port, the Bengawan Solo was rechannelled so that it emptied directly into the Java Sea. It was an engineering feat: twenty kilometres of straight channel undertaken because discharging the river's sediment into Madura Strait was threatening to make it impassable.

Such changes in the course of rivers — engineered by humans or natural — have occurred many times, and we'll consider what this means in terms of rising sea levels shortly. But perhaps more important to the overall story of climate change is what can be seen from a Google Earth shot. It suggests that one of the main reasons for the expansion of the port facility at Gresik is that farther south in Surabaya, there just simply isn't much room to expand. Keeping the world economy rolling with shipments of chemicals and other products is the equivalent today of the spice trade that made the Ocean Isles so spectacularly important for centuries.

A warning, though, to all who think they can cash in on sudden increases in demand for certain goods: neither Gresik nor Surabaya is protected from the effects of bad weather and rising seas. Just a week or so after the announcement of the new installations at Gresik, floods and storm surges hit communities along the coast. Flash floods roared through the immediate hinterland, damaging at least two hundred homes, flooding roads, and inundating 575 hectares of fish ponds.[28] Disaster for some, and all this against the background of Covid-19.

~

Meanwhile back in Jakarta, what was the outlook for Ramadan?

> The government has banned the Idul Fitri tradition of *mudik* (exodus) for the second consecutive year to curb the spread of COVID-19, which has spiked after major public holidays in the past.
>
> Coordinating Human Development and Culture Minister Muhadjir Effendy said the ban intended to prevent a spike in COVID-19 transmission and deaths after the holiday. Such an increase was observed early this year, after Christmas and New Year.
>
> "Following a coordination meeting among ministers on March 23, it was decided that the government will ban *mudik* in 2021," said Muhadjir during a virtual press briefing on Friday. "The ban is effective for civil servants, the military, the police and the general public." ...

Under normal circumstances, some 20 million people in Greater Jakarta travel to their hometowns during the holiday.

Muhadjir said the ban would be in effect from May 6 to 17. He urged people to avoid nonessential travel outside of their cities of residence during the period to prevent a spike in COVID-19 cases.

As of Friday, health authorities had recorded nearly 1.5 million cumulative COVID-19 cases and about 124,000 active cases. The weekly test positivity rate — the number of cases detected out of the number of tests conducted — stood at about 13 percent on Friday.[29]

The story — "'Mudik' Banned Again" — in the English-language *Jakarta Post* on March 26, 2021, was accompanied by a photograph of people streaming across the Suramadu Bridge, which crosses Madura Strait at Surabaya, during Ramadan in 2020. Despite rising concern about Covid-19 at that time, thousands and thousands of Indonesians went home for the holidays, many carrying the virus with them.

By that time, the Indonesian government had quit talking about its plans for moving the administrative capital from Jakarta. Securing vaccine supplies, coping with flooding throughout the country, and dealing with the Covid dead were taking up much energy.

And on the other side of the world, the paintings of rainbows that my grandkids had made at the beginning of the pandemic were beginning to look a little faded by March 2021. My husband and I had our first vaccination shots, but the disease was still swirling around out there. The hope of visiting Jakarta and seeing first-hand what the city looked like was becoming dimmer and dimmer. But — perhaps ironically — how people were reacting to this rolling crisis was producing hints of how we'll all be able to deal with rising seas.

To return to that photograph: the crowds crossing on foot would walk more than five kilometres across the longest bridge in Indonesia. The island they were going to is low and dry and very, very poor. Probably its biggest product is salt — indeed, it's frequently called the Island of Salt — which suggests just how its climate differs from that of the well-watered western parts of its neighbour, Java. Lying in the rain shadow of Java's volcanoes, it does not receive the kind of precipitation that is necessary for the lush rice fields that have fed millions for centuries. This means that "salt farmers" can allow seawater to dry in enclosures that bear a strong resemblance to the famous Dutch polders, producing salt crystals after a few weeks. But that, cattle raising, some fishing, and subsistence farming is not enough to support a population, and over the twentieth century, many Madurese left to find work elsewhere. As if to underscore the sorry conditions, forty-six pilot whales were stranded off a beach on the south side of the island shortly before the 2021 pilgrimage of Madurese. While islanders tried to save them, only three were persuaded to go back out to sea.

Just another chapter in these woeful times, I thought as I pored over the Google Earth shots of Madura and then compared it with Java, wondering what the draw must be for people to walk home for the holidays.

The answer is that home is home, of course. In the larger context, preserving home is what people have wanted to do over millennia as the waters rose.

4

WRESTING A HOME FROM THE SEA

The statue of Evangeline stood in front of the church-turned-museum on that gorgeous early June day. The name rang a bell with me: my grandmother — she who told the story about the little boy who saved the Netherlands by putting his finger in a break in the dike — could recite verse after verse of Henry Wadsworth Longfellow's poem about her. She did it so often that even I remember bits:

Prelude

This is the forest primeval. The murmuring pines and
  the hemlocks,
Bearded with moss, and in garments green, indistinct
  in the twilight,
Stand like Druids of eld, with voices sad and prophetic,
Stand like harpers hoar, with beards that rest on their
  bosoms.
Loud from its rocky caverns, the deep-voiced
  neighboring ocean
Speaks, and in accents disconsolate answers the wail
  of the forest.

• • •

> Ye who believe in affection that hopes, and endures,
>   and is patient,
> Ye who believe in the beauty and strength of woman's
>   devotion,
> List to the mournful tradition still sung by the pines of
>   the forest;
> List to a Tale of Love in Acadie, home of the happy.

The poem was a story, she told me, of a young woman who had been forced from her country and then wandered all her life looking for the young man with whom she had been in love. Beautiful sentiments, lovely poetry, enough for young girls to cry over, she suggested. And I suppose she did, back at the turn of the twentieth century when the poem was taught all over the United States.

Be that as it may, there was a disconnect between the slim and lovely girl the statue represented, the poem, and the landscape I saw around me. No murmuring pines, no rocky caverns, and the sea, which wasn't far away, I knew, was hardly echoing disconsolately the wail of the forests. What I saw on this late spring morning were rolling fields and flowering apple trees. Green grass. A neat church that probably dated from around the time of my grandmother's youth. There was a new exposition hall — modern boxy lines, brick and concrete — for this place is both a Canadian National Historical Site and a UNESCO World Heritage Site.

We hadn't been looking for much more than a pleasant drive around Nova Scotia after spending a couple of days in Halifax for some meetings. We were hoping we'd be able to have lobster or some other succulent seafood for supper at a little restaurant looking out over the Bay of Fundy that evening. But it was lunchtime, and the Grand-Pré National Historic Site looked like just the place to eat the lunch we'd brought with us.

Perhaps you know the story of the Expulsion of the Acadians from Nova Scotia. Unlike the poem, it's taught in Canadian high schools frequently, and it may occasionally be mentioned in American history classes. Some of the Acadians, after all, ended up in Louisiana, where they became the

Cajuns, whose colourful culture rocks Mardi Gras celebrations each year. The Expulsion of the Acadians — or *Le grand dérangement* — happened in the 1700s, when the British were trying to wrest this territory from the French as part of their long, long conflict with that other pretender to control of North America. Through twists of diplomacy and warfare, the British were poised to take over this very fertile countryside, but they insisted that the French-speaking farmers sign oaths of loyalty to the British crown. Many wouldn't do that, not necessarily because they had much desire to be French (their ancestors had been farming here for more than one hundred years, and the old country was far away) but because they were fiercely independent and proud of what they had wrought from the salt marshes on a bay that has the highest tidal variation of anywhere in the world.

Because of their steadfast refusal, many of them were evicted, a drama that is not forgotten in Acadie (although Evangeline is more of an American myth than an Acadian one). The amazing transformation they had wrought on the tidelands was almost destroyed, but the few who remained, and the other people who came from elsewhere but learned from them, succeeded in preserving the lands they had reclaimed from the sea.

What the settlers from France did was very much like what the Dutch were perfecting at about the same time; that is, building a system of sluices and dikes that kept out water at high tides and allowed fresh water to drain into the sea at low tides.[1] The first attempt, called Port-Royal, was made in 1605 when explorer Samuel de Champlain and his companions put up a settlement of wooden buildings not far from where the Annapolis River empties into tidewater. Note the date: Batavia on Java was founded fourteen years later and the Dutch settlement that became New York City was laid out in the mid-1600s on a plan that resembled the original Amsterdam. Traces of it can still be seen on Manhattan where Broadway follows the line of the canal built by the Dutch nearly four hundred years ago.

Port-Royal wasn't successful, but an attempt a few decades later farther up the Bay of Fundy thrived. The first polders were relatively small, but by using the sods cut from the deep-rooted grass of the salt meadows to build dikes, the settlers at Grand-Pré were able to transform the marshlands rather quickly into fertile fields. The soil had been enriched over hundreds of years with silt left by tides and floods; all that was needed was to control the water budget so that rain washed the salt from the land and sluice gates kept out seawater at high tide. Within a couple of seasons, the meadows had become arable land, at much less effort than it would have taken to cut down the trees growing in the thinner soil of the higher ground. Within decades, hundreds of hectares of marsh had been transformed into the *grand pré*, the big meadow. Even though building the dikes meant cutting away some of the sod to build up the dikes' walls, the spring floods regularly left a load of silt to replenish the land.

Keeping these dikes and the sluices — called *aboiteaux* — in good repair meant constant surveillance and coordinated community work. Over time, the farmers perfected special tools for cutting the sods that were used to repair the dikes. According to interviews gathered at the turn of the twentieth century, the sharp-edged spades used to cut quickly and cleanly through the strong tangle of grass roots were kept ground to a knife edge and hanging in outbuildings where they could be grabbed quickly when an especially high storm surge overtopped the dikes. Children weren't to touch them, one of the elderly farmers remembered.[2]

Fair enough — I'm reminded of one small boy I knew very well who was sure that a sharp knife was going to jump out and cut him — but the question arises: Where did these settlers learn how to take the land from the sea?

As I said before, this work was being done about the same time that the Dutch were perfecting their polders, but closer examination of the records from the period suggests that rather than using the Dutch techniques — which were pretty much state of the art — the Acadians took some other ideas from the Atlantic coast of France and refined them. Many of the *colons*, or settlers, came from the La Rochelle region, north of where the Loire River forms an estuary and then enters the Atlantic. Farmers there had been drying out swamps for ages and protecting them from rampaging river water, so it figures that many in those first groups had some idea of what to do.

In addition, among the crew that settled at Grand-Pré, there apparently were several men who were skilled in making salt, which was an important commodity used, among other ways, to preserve the cod that packed the ocean's waters. In the long run, they must have done well for themselves, to judge by the many people who today bear the family names *Saulnier* (salt extractors) or *Saunier* (salt merchant), like ice hockey player Jill Saulnier and Quebec politician Lucien Saulnier. But unlike in parts of France and on the island of Madura in Indonesia, making salt wasn't easy around the Bay of Fundy because the summers were frequently short and rainy. The result was that the salt-making attempt was stillborn, but the skills needed to build the holding ponds were just what was needed to construct the dikes and sluices for transforming the salt marshes into meadows.

Present-day Acadians are justly proud of what their ancestors accomplished, but while their great meadows are wonderful, the Acadians weren't the only newcomers who decided that salt marshes were much better than high ground when it came to starting their life on the North American continent. About the same time that the Acadians were beginning their monumental project, English, Dutch, and Swedish colonists were building dikes and sluices along the eastern seaboard of what is now the United Sates.[3]

It should be noted that more than seven hundred thousand acres of fens — swampy land in southeast England that might once have been connected to Doggerland — had also been reclaimed with dikes, sluices, and windmills by the mid-seventeenth century. That means that in addition to the settlers from France, those from parts of England knew what could be done with swampy land. As M.J. Harvey imagined in an article in *Nature Canada*, suppose you were looking for a good place to claim as home. Where would be the best place to settle? Not on the densely forested uplands, because it would take a man years to clear a reasonable farmstead. What you would want was a place with good pastures and with deep soil, free of stones and gravel. The great tidal salt marshes along the shores and lining the estuaries were just what was needed because they could be used almost immediately.[4]

Indeed, as Sherman Bleakney notes, it would have been silly to try to scratch a living from wooded upland when the salt marshes were there to be tamed. Later, some Anglophone settlers scoffed at what the Acadians had accomplished, but what they did made great sense. Of course, building the dikes and sluices required a sense of community action, of the group working for everyone, which might not have appealed to individualistic folks who just wanted to stake their claim and were ready to go it alone in the hardscrabble uplands.

A couple of hundred years later, residents of the St. Lawrence valley found themselves faced with the same sort of situation. By then, the most fertile land in the valley had come under cultivation. The farmers had prospered, providing plenty of food necessary for very large families to thrive; travellers in the early nineteenth century commented on how tall and strong the *habitants* in Quebec were. But after a generation or two, the farmsteads could not support the new families that inevitably formed as the population grew. By the middle of the nineteenth century, it seemed that only two possibilities were available for young people looking for a way to make a living. One was to head to Montreal or down into the "Boston States" to find work in the new factories and mills that industrialization was spawning. The other, encouraged by a segment of the Roman Catholic clergy, was to clear more land in the uplands, away from the river, to colonize the country and by so doing save the French language and preserve Roman Catholic values from outside influences. A whole literature has been written about these population movements. In fact, there was a third possibility, one that has received less attention. This was a movement to transform the salt marshes along the St. Lawrence. It was championed by a new agricultural college in the Kamouraska region, the heart of the marsh country.

Established in 1859, the institution at Sainte-Anne-de-la-Pocatière was only the second agricultural college in North America, and for several decades it was on the cutting edge of a scientific approach to agriculture. This led to a melding of old and new techniques. Dozens of *aboiteaux* were constructed over the following decades, with thousands of acres transformed from tidal lands to agricultural production. One of the main crops was excellent salt hay. It was used to feed cows that produced great quantities of milk from which high-quality butter was made for markets up and down the St. Lawrence and into New England. The hay was also shipped out to feed the horses that powered so much of life at the time.

But the story doesn't end there. In the twentieth century, many more hectares of tidelands were diked off, so that to drive through the Kamouraska region in summer now is to travel though flat green countryside promising plenitude. Though the great period of *aboiteau* building ended in the 1980s, the fertile fields remain. We saw that first-hand one hot and sunny August day in 2021 near the pretty village of Saint-André-de-Kamouraska, some 425 kilometres down the St. Lawrence River from Montreal.

It was nearly noon when we stepped out of the car and the scent of the clover in the fields on either side of the parking lot enveloped us. Honey, I thought; the air was almost as thick, liquid, and scented as clover honey. Purple flower heads contrasted wonderfully with green leaves. The fields spread out over the flat land until they ran into the tall granite outcropping that hinted at the geology of the region, the ancient bones of the Earth. These monadnocks are the hard cores of truly ancient volcanic intrusions that resisted the scouring caused by moving ice sheets. The river's estuary (one of the largest in the world, by the way — by some measures even bigger than that of the Ganges-Brahmaputra) is relatively shallow along here. The main shipping channel is on the north side, where the depths are greater.

The waters of the St. Lawrence estuary couldn't be seen from the parking lot. They lay just on the other side of the embankment up ahead. Nor did I notice at first the channels that ran through the fields, designed to carry rainwater and snowmelt toward the tide flats. Once we crossed the gangway that connected the parking lot to the embankment — called a *batture* — the lay of the land became literally more apparent.

The tide was out, and a kilometre or so of the salt marsh was visible, as it has been twice a day for thousands of years. For centuries First Nations people spent much time along this coast, fishing, hunting, and gathering various useful plants. Kamouraska, the region's name, comes from what those first inhabitants called it: a place of rushes by the water.[5] On the dry-land side of the *batture*, a small brook meandered through what looked like an ordinary stream bed — it hadn't rained in a week or so, and obviously there wasn't much water to be drained from the fields. On the seaward side of the *batture*, however, a channel filled with tidewater ran straight toward the great expanse of the river. Connecting the two sides was the *aboiteau* it-self, a concrete tunnel with two gates that were closed at the moment. When high tides came rushing in, the pressure of the incoming water would hold them shut. But after a storm or in the spring freshet season, the fresh water

A channel in the salt marsh drains water from the *aboiteau*.

that collected in the fields would force the gates open from the other way, so it could run toward the sea at low tide. The set-up was very much like the one the Acadians had developed some three hundred years before, only modern materials were used, not the hollowed-out logs that once were the conduit for the fresh water.

These *battures* and *aboiteaux* — along this stretch the *batture* extended about eighteen kilometres and there were several sluices — require considerable maintenance. A big winter storm in 2010 (about which more later) sent waves over the top in several places, and parts of the *batture* had to be reinforced and made higher. Now wild rose bushes and salt-resistant trees grow along the top, their roots helping to consolidate the dike. Not nearly as big as the dikes that protect the north of the Netherlands from the sea, the *battures* here and elsewhere along the coast are nevertheless impressive constructions.

The *aboiteau* or sluice through which water drains.

But it's also important to note that there's been a switch in the way some observers are thinking about marshlands and tide flats like those along the St. Lawrence, and this accounts to some extent for the end of building new *aboiteaux* in this region. Transforming tidelands into meadows was considered a collaborative effort between humans and Nature for a good part of the nineteenth century and well into the twentieth. It was thought that most marshes were actually growing as increasing amounts of sediment were deposited, and that transforming them was really "improving" them. It's now apparent, though, that human activity is playing a much larger role in what is happening at the water's edge. Where the shoreline marshes appear to be growing, it probably is due to more silt being carried downstream because of deforestation and other damage to the land. But the slow rise of seawater will eventually endanger the dikes as the salt marshes that now muffle the waves are eaten away.

The fields drained by the *aboiteau* are actually lower than the salt marsh.

The *batture* is the raised dike that keeps back salty water from the St. Lawrence estuary.

Which brings us back to the vagaries of rivers and the dangers of rising waters. Let's look at three places where water and land come together and people must make changes.

To quote the Chinese sage of the sixth century BCE, Lao Tzu, "There is nothing in the world more soft and weak than water, yet for attacking things that are hard and strong there is nothing that surpasses it, nothing that can take its place."[6]

PART 3

RIVERS, DELTAS, AND ESTUARIES: THE BATTLE BETWEEN WATER AND LAND — THREE EXAMPLES

MUSICAL INTERLUDE

Let us pause for another musical interlude, this one about a river. Until I started working on this book, rivers had been simply part of the scenery of my life, even though for decades I've lived almost within walking distance of one of the greatest in the world, the St. Lawrence. Montreal is an island, but people who live there these days find it easy to forget that; there are bridges over the river and tunnels under it, and with very few exceptions, it is well-behaved. Oh, there have been ice jams occasionally in the spring, and at the beginning of the city's history, one caused a historic flood. More frequently, the smaller river that runs around the north of the island misbehaves and overflows onto surrounding lowland.

Lately though, as I stand beside the rapids that blocked navigation upriver before canals were built around them in the nineteenth century, I have begun to hear in my head certain phrases from Bedřich Smetana's tone poem "Vltava" (or "The Moldau"). The piece is part of a long suite of music that the patriotic Czech wrote about his homeland. It begins with two flutes gently evoking the sound of two small streams that have risen in mountains. They roll along independently at first and then come together. Other wind instruments join in as the water rushes downhill. A triangle sounds three times, and a new theme is introduced, played by the full orchestra. It recurs again and again, as the music and the water become one and roll onward. Percussion and horns signal the river's tumultuous passage through rapids before it arrives at its destination, where the theme resolves the piece in a final burst of energy from the entire orchestra.

A good performance can be found on the Kennedy Center's YouTube channel.[1] The National Symphony Orchestra plays before an enthusiastic audience of young people, who applaud when they like what they hear and not at the end of the piece as seasoned concertgoers would. This delightful break with tradition demonstrates the power of one aspect of human civilization, that of music, which hopefully will survive the rising waters.

5

SHANGHAI

The St. Lawrence drains about a quarter of North America; only the combined drainage basins of the Mississippi and Missouri have a larger area. Looked at one way, the St. Lawrence has its origins in the Ice Age, because the Great Lakes, those immense bodies of fresh water that filled the hollows left when the glaciers melted, would not be there had it not been for the periods of glaciations of the past. But on the other side of the world, there are a half-dozen rivers that are even more wedded to the remnants of that faraway time, since they flow from the icefields still remaining in the Himalayas and the Hindu Kush. As we consider what rising seas may mean, it is instructive to look at them. We'll discuss the duo of the Ganges and the Brahmaputra-Jamuna, which rise in different parts of the great mountain ranges but finally mingle their waters in the Bay of Bengal, where Dhaka, one of the region's largest cities, could be washed off the map in a few years by rising seas. But before that, let us turn to the longest river in Asia, the third-longest river in the world, the Yangtze, and the city located where it joins the East China Sea.

By some accounts, Shanghai is the world's most populous city. Officially it now has twenty-four million permanent residents, but it also is a temporary home for migrant workers who have come from other parts of China to build its skyscrapers and impressive transit system and to remake the very land it stands on. *Shanghai* means "on the sea," and when it began to grow from a village of fisherfolk in the eleventh century, it really was on the coast.[2] At the time, it wasn't much of a settlement, cut off as it was from one of the main channels of transportation in the Middle Kingdom, the Grand Canal.

For millennia China has had two main axes of development, the Yangtze and the Yellow River to the north. Both provided water for agriculture and nurtured the deep and rich civilization that began developing in the centuries before the Common Era. But communication from north to south, from river valley to river valley, was difficult. The first canals designed to correct that date back to the time poetically called the Spring and Autumn Period (770–476 BCE). One of the local powers then had its seat at Suzhou, west of the current Shanghai. In order to transport grain and other goods, the king of the day ordered the construction of waterways to connect rivers and lakes so that supplies could be sent north. This was the precursor of the aptly named Grand Canal.

Even though it was used extensively for centuries, there were periods when the canal system's maintenance was neglected and parts silted up. Then, as Western Europe floundered in the depths of the Early Middle Ages after the collapse of Rome, construction began to revive and extend the canal, which Jacques Gernet describes in *A History of Chinese Civilization* as "the world's largest and most extensive civil engineering project prior to the Industrial Revolution.... By the 13th century it consisted of more than 2,000 km of artificial waterways, linking five of China's main river basins."

The Grand Canal was where the action was when Marco Polo made his journey into China, and Shanghai didn't rate a visit. It was just a small settlement on the shore, removed from the real economic and cultural activity. Polo was impressed by what he saw in the region, however, calling Hangzhou, the canal's southern terminus about 160 kilometres southwest of Shanghai, "without doubt the finest and most splendid city in the world." He says the city had a circumference of 160 kilometres with a lake on one side and a big river on the other, and that "many channels, diffused throughout the city, [carry] away all its filth." And, to give a measure of its economic activity, he quoted customs officials on the amount of pepper used to season the city's food each day: "43 cart-loads, each cart-load consisting of 223 lb."

Shanghai, although still a village, had already benefitted from the technical prowess of Chinese engineers. Around 238 BCE, Lord Chunshen ordered the construction of a channel to carry water from a lake south and west of Shanghai to the Yangtze. The aim was to improve water-borne traffic and to lower the risk of flooding in the low-lying country, which was already a problem. The result was the Huangpu River that flows through Shanghai today.

Along its 113-kilometre course, the Huangpu is fed by several streams, and now when it passes through Shanghai it is three-quarters of a kilometre at its widest, carrying enough water to provide most of the city's water supply until recently.[3] However, the river is not really tamed, despite levees constructed along its course — the most famous is the Bund on its western side, along which European and American interests built grand art deco buildings in the 1920s and 1930s. The river still floods occasionally, despite the Bund's levees, which are 6.9 metres high, and the tide gates that control flow from the Suzhou Creek, a stream that enters the Huangpu in the northern part of Shanghai.[4]

The Bund holds back the waters of the Huangpu and is also a grand promenade.

The long history of water management in China demonstrates that the Chinese have no fear of bringing out heavy artillery to solve problems. How that has played out in Shanghai is complicated. Take the case of the huge load of sediment that the Yangtze carries. (The Yellow River does too, by the way — it gets its name from the yellow soil it carries to the sea.) Because of the silt, the Yangtze delta has edged seaward many kilometres over the last thousand years through the accumulation of sediment, and today the Huangpu must be dredged continually to keep its channel open where it enters the Yangtze.[5]

At first, the delta land east of the Huangpu was not good for agriculture, just as the salt marshes in Acadia and along the Kamouraska shore had to wait until the rains had cleansed the fields their *aboiteaux* protected. But, as in other places, the marshland was very fertile once the salt leached out, and farmers discovered that cotton grew well there. A machine to remove seeds from the cotton bolls made its appearance around 1314, almost five hundred years before the American Eli Whitney invented his cotton gin. The machine's effect was similar to that of Whitney's machine, and cotton fibre and cloth production became much easier and increased rapidly. As the quality and quantity of cotton cloth grew, Shanghai became a major trading centre for both the finished product and the soybean cake fertilizer that was imported from the north to enrich the cotton fields.[6] Trade in silk and cotton quickly became important. Peasants in the region concentrated on growing cash crops like cotton rather than on producing food for themselves.

Shanghai's growing prosperity was not without its problems, since it attracted the attention of Japanese pirates who prowled the coast and the waterways draining into the sea. About the time Marco Polo started back to Europe, five villages on the Huangpu were consolidated into a municipal administrative unit of Shanghai, in part to provide a united front to interlopers. Note that today, Shanghai, sprawling over more than 6,335 square kilometres, is much, much larger than the original unit, as the surrounding area is now included in the municipality, which is also the most populous in the country.

In 1553 the town elders petitioned for permission to build a wall to protect the city from marauders. Its outline can still be seen today in the circular arrangement of streets not far from the Huangpu in the Puxi district. When the wall was torn down in 1911, Shanghai was bursting at the seams, and the international community, attracted by the trade that flowed through the port, had acquired rights to build their own settlements outside the wall, which were quite independent of the larger Chinese administrative structure. (Another one of Shanghai's ironies is that the building where the First National Congress of the Chinese Communist Party was held was in this international area. French Concession police, tipped off that a revolutionary meeting was being held, scoured the neighbourhood for several days before they found it. The delegates fled and then had to continue their meeting on a pleasure boat on a lake about one hundred kilometres out of town.)

When I visited Shanghai in 2005, the city was preparing for a world's fair to be held five years later with the theme "Better City, Better Life." At the time, it sounded very engaging and right in line with the book I was working on then, *Green City: People, Nature and Urban Places.* My idea was to look at the way people in various cities brought Nature into their lives, either formally through parks, public landscaping, and street trees, or informally in gardens big, small, and balcony-sized. I also was interested in how some cities were able to integrate agriculture into urban life.

Shanghai did not disappoint me. Officials told me then that their goal was to make 35 percent of territory "green," providing fifteen square metres of green space per resident by the time the world expo opened. As I travelled along the elevated highways into Shanghai on the airport bus, this green effort was already evident. Steel mills and industrial plants lined the edges of the nearby waterways, their red-brick buildings smudged by smoke, with grey and black piles of slag and other waste lining the surface roads. But the edges of several compounds were planted in bushes and trees, producing a green that contrasted brightly with the dark industrial tailings.

The highway rights-of-way were also lined with green, with footpaths and benches that people, at least in the centre city, used like any other park. I later saw, farther out in the new towns, that district governments often made other choices, grouping the required green spaces together to produce big parks filled with sports facilities.

The Huangpu runs roughly from south to north through Shanghai, dividing Puxi, the older part of the city on the west side, from Pudong, on the east. Formerly, Pudong was a district of industry, slums, and small farms, but since the 1990s it has profoundly changed. The part of Pudong directly across from Puxi and Shanghai's older financial and commercial centre is now a forest of extreme high-rises, new industry, and homes for hundreds of thousands of people. Several other formerly rural districts have been incorporated into what is now called the Pudong New Area, which is governed as part of greater Shanghai. Not only has the population and economic activity soared, the actual land area has been considerably increased as the sediment washed down from the interior of China by the Yangtze has been used to form new land. So much has been added that Shanghai now boasts a larger area of reclaimed land than any other city in the world. Between the mid-1980s and 2017, Shanghai as a whole actually increased in area by 6.46 percent because of the additions to Pudong; the reclaimed area amounted to 585 square kilometres of new land.[7] (During the same period Jakarta added eighteen square kilometres or 4.18 percent of its area, putting it in fourth place in a study of sixteen megacities. We'll look at the implications of that later on.)

≈

Shanghai has, for a long time, been a study in contrasts, and the one that exists between today's city and the city that existed before the Communist Revolution could not be starker. China's pride in the city's — and country's — current prosperity explains a lot about the concerted effort being made to harden the coast against rising sea levels.

For decades in the early twentieth century, Shanghai was one of the most glamorous cities in the world, with temples of international finance, high-stakes gamblers, thriving industry, and a large, wealthy foreign presence. Alongside that, though, existed a huge underclass. The poor, often from rural areas, were everywhere as a war for control of the country by home-grown forces raged from about 1927; the invasion by the Japanese in 1937 only interrupted it.

The British novelist J.G. Ballard, who spent his childhood there in the 1930s and during the Second World War, was marked by the contrast. He gives a poignant chronicle of this period in *Empire of the Sun*, the tale of a boy's life in war-torn Shanghai. One Sunday in December 1941, he writes, when Jim, the hero, is leaving for a party in the family's big Packard, the family's driver runs over the foot of a beggar who had taken up residence two months before just outside the entrance to Jim's home.

But the man — "a bundle of living rags whose only possessions were a frayed paper mat and an empty Craven A tin which he shook at passers-by" — does not move from his mat, just as he had refused to move from this place when other beggars challenged him. His position does not make it easier for him in that winter of hard times, however, and after a week-long cold spell, he is too tired to raise his tin. Then, the morning after a heavy snowfall, the boy sees that the man is covered in snow, with his face peeking out like that of a child snuggled into bed under a thick white quilt. But rather than be alarmed, Jim tells himself that the man doesn't move "because he [is] warm under the snow."

More hard times follow. Jim, who is old enough to ramble around on his own and young enough to accept without questioning the differences between the life of ordinary folk and his privileged one, recounts what he sees even when his family is interned in a Japanese concentration camp for foreigners.

The novel ends with the war's conclusion, but that was hardly the end of the wildly different paths that the different levels of Shanghai society took. Civil conflict continued after the Japanese defeat, ending only with the Communist Revolution in 1949. There followed nearly a half-century of struggle, which included a massive famine in the 1950s and ten years of the disastrous Cultural Revolution in the 1960s and '70s. Only after the death of Mao Zedong in 1976 did China's leaders change economic and cultural tactics. Even then, the path was never a straight one upward — witness the repression that followed Tiananmen Square protests in 1989 — but by the end of the twentieth century, China had become an economic powerhouse with vast ambitions.[8]

~

Among the signal objectives of this new China was the colossal project of rehousing the nation by 2020, which resulted in China using as much concrete between 2010 and 2013 as the United States had during the entire twentieth century. The effect on climate change of this initiative has been considerable, since CO_2 emissions from the production of cement, the major ingredient in concrete, account for between 5 and 8 percent of total annual CO_2 emissions. Then there is the way of life these new developments allow. More roads mean more cars, particularly in a country like China, where the population wants to catch up with the consumer standards of the West.[9]

When I was in Shanghai, I got a taste of the immense transformation the Chinese were embarking on when I took the Metro from Puxi, where I was staying not far from the French Concession, to Longyang Road Station, which was the end of the line at that point. Nearby was the Zhangjiang Hi-Tech Park, a cluster of research and technological institutions, where many buildings were surrounded by fences through which one could see some pretty nice landscaping. But much of the surrounding area was raw still and reminded me of developments outside other cities that I know — lots of construction equipment and land scraped clean of vegetation.

As I said, I had taken the Metro to the end of the line, but now it goes much farther, all the way to Shanghai Pudong International Airport, whose runways abut the sea. Arriving or taking off, you can see just how flat and undeveloped much of Pudong still is. There have been remarkable changes in the last few years, however. Recently, I spent an afternoon comparing the maps I'd brought back from my trip with current ones, and it was clear that not only had a lot been built, to judge from the number of new roads featured in the more up-to-date maps, but also in some places the landscape was completely changed. Part of this new land is dedicated to port facilities and a new free-trade zone, but much of it is designed to attract hundreds of thousands of residents as part of Shanghai's plan to create five new satellite sub-cities to reduce population pressure on central Shanghai.

Chief among these is the Nanhui New City (formerly Lingang New City) built in Nanhui district, where the Pudong peninsula noses into the sea north of Hangzhou Bay. It is about seventy kilometres from downtown Shanghai and a good sixty kilometres from where the Metro line ended when I was there. Even then, though, planning was underway for the new urban centre. The idea was to build a city that would be considerably different from older Chinese cities, as well as distinct from the many new cities that were then being designed in China's effort to rehouse three hundred million people by 2020. At the turn of the twenty-first century, the German urban planning firm GMP won a competition to lay out the new city. Rather than concentrate density in the centre of the new city, as is the case in traditional European and Chinese cities, the firm planned a circular lake as the centre, with roadways and neighbourhoods radiating out like "concentric waves, produced by a drop hitting the water surface."[10] What the urban planners called wedges of greenery would cut through the new city too, which would also feature watercourses and small lakes. These last would contribute to making Lingang a "sponge city," an urban area where water can easily percolate back into the earth in an attempt to cut down on flooding in times of heavy rainfall.[11]

One of the main features of the new urban centre is the Yangshan Deep-Water Port, which while not an actual part of the city, is nevertheless one of its raisons d'être. The sea off Pudong is shallow enough to make dredging up sediment to increase the land area relatively easy, but Hangzhou Bay to the south has deep-water sections. The deep-water port, with tens of docks for unloading container ships, is connected to land by the longest marine bridge in the world. Constructed on two islands that rise steeply out of the bay, the port complex is one of the biggest ones around and surpassed that in Singapore in size and action in 2010. Videos of traffic on the 32.5-kilometre bridge show a steady stream of trucks, but apparently, despite traffic, the greater ease of unloading and the larger area of the port makes it the preferred choice for many carriers shipping into Shanghai, because the megacity's other port installations on the Huangpu and Yangtze Rivers require navigating crowded ship channels. Furthermore, the Chinese government has plans to make Lingang and the surrounding area in Pudong home to many new industries catering to overseas markets. Using the bustling Yangshan Port, which is also in a free-trade zone, can considerably reduce factory-to-ship time as well as offering advantages such as preferential customs handling.

But can all this development stand up to the menace of rising sea levels and the threat of ever stronger storms? It is worth noting that the Yangshan Port facilities are all built on the western, more protected side of the islands in Hangzhou Bay. The high hills of the islands may give the anchored ships and the waiting containers some shelter when push literally comes to shove during major storms.[12]

From the time that intensive development began in what is now the Pudong New Area, environmentalists questioned the wisdom of the dredging required to increase the available land. Back in 2010, Liu Zhengdong, then deputy director of the East Oceanic Division of the State Oceanic Administration, told the *Global Times*, "The ecological system of our ocean has already been severely deteriorated due to over reclamation from the sea. Diversity of species has decreased, and lots of wetlands are disappearing." But Zhao Youcai, a professor from the College of Environmental Science and Engineering of Tongji University, countered by saying that the costs were worth the benefits: "Since the new facility built on the reclaimed land is expected to bring huge economic benefits, the government can ignore possible environmental problems that it may pose to the city."[13]

Much of the criticism of land reclamation has focused on the destruction it wreaks on wetlands, which are important for birds and fish, but they have an important role to play in mitigating rising sea levels, too. As Ren Wenwei of the World Wide Fund for Nature told *Scientific American*, wetlands "also [serve] as a safeguard for the city's … residents."[14] (This is a point we'll return to later, you can be sure.)

Obviously, the pro-development position carried the day. But there are studies currently warning that by 2030, Shanghai, like Jakarta and so many other cities, will be experiencing serious problems due to sea-level rising and increasingly violent storms.[15] Currently, the average elevation in greater Shanghai is only four metres above mean tide level, which for the last fifty years has already been rising at a rate of about 1.6 millimetres annually. Furthermore, because it is constructed on alluvial soil, the very ground on which the city is built is being compacted, which means more sinking.

Even worse, flood waters from the lake that is the source of the Huangpu can raise the level of that river disastrously and quickly. During the 1990s, flood-water levels on the Huangpu rose to five metres five times, but in 2001 alone this happened four times. Shanghai is protected to some extent by a levee along the Huangpu designed in the 1950s to withstand a one-hundred-year flood. It was upgraded following Typhoon Winnie in 1997, when the Huangpu levees were breached and four hundred homes were flooded. What the future holds is unknown, of course.

But it's not just the venerable Huangpu whose water threatens the city. Seawalls designed to withstand level-12 typhoons protect 520 kilometres of shoreline around the city. They run along the south bank of the Yangtze downstream from where the Huangpu enters it, along the north bank of Hangzhou Bay, and around Changxing Island, which contains the reservoir from which most of the great city's tap water now comes. Just how well these sea defences will work ten or fifteen years from now remains to be seen. Not surprisingly, Shanghai is considering constructing a tidal barrier on the Yangtze like the one that protects London from storm surges. The British engineering consulting firm HR Wallingford has developed a plan to build one, as well as to raise dikes and levees along the Huangpu by 1.1 metres.

Deciding how big and where to build the barrier are daunting problems. One of the key factors that needs to be considered is how frequently the tidal barrier might be closed. If doing that is required frequently, the time for maintenance is reduced, so the risk of failure increases. Navigation to Shanghai's port facilities would also be affected by frequent closures. Wallingford considered a system that would include locks to allow traffic in and out of the river. It also factored in the problem of flooding from upstream when the Huangpu and the Yangtze roar after heavy rain. The firm notes, almost with surprise, that the solutions they found most promising resemble mightily the Thames Barrier that protects London, and to which we'll return later. Just when the Chinese will have a Shanghai version in place hasn't been determined yet.

The embankments currently protecting Shanghai are only part of China's "hard" defences against the sea. The seawalls and other barriers are now longer than the ancient Great Wall, lining more than 60 percent of the 14,500-kilometre total length of the coastline.[16] To be sure, there are a few projects to restore wetlands, but they are very small when compared to the length of the "new Great Wall." In February 2021, the *China Daily* reported proudly that a 47.33-kilometre stretch of coastline around the Bohai Sea had been restored, but that is minuscule compared to the length of the armoured shore.

That the Chinese are capable of doing marvellous things to manage water in nearly all situations, and are willing to do so, is not at question. One current example of such developments is the fresh-water reservoir, mentioned earlier, built on Changxing Island in the middle of the Yangtze, just opposite the point where the Huangpu enters the bigger river. The nearly seventy-square-kilometre reservoir, built on the north side of the island, was opened in 2010. It is filled by water pumped from deep in the Yangtze, which is considerably less polluted than the surface water. It has a reservoir with a capacity of 430 million cubic metres and supplies about 70 percent of Shanghai's tap water.[17]

Flood warning systems and systematic cleaning of storm drains are also now part of Shanghai's disaster plan, while an increasing amount of environmental sponge-city-type flood control is being put in place. The proposals sound quite a bit like what is being proposed for the "normalization" of Jakarta's flood-prone rivers. The difference is that here the idea is being given a trial instead of just being talked about.

Will all this work? The answer isn't clear. Early summer 2020 saw weeks of rain in southern China, which swelled the region's rivers and produced some of the worst flooding in years. The long Yangtze River basin was particularly hard hit, with water filling the massive Three Gorges Dam to near overflowing. Built to generate hydroelectricity, but also to head off floods, the dam faced its biggest challenge since it opened in 2003. It held, but sluice gates had to be opened in order to reduce pressure from the river water. As a result, several areas adjacent to the river were flooded. Wuhan, located about 690 kilometres west of Shanghai on the Yangtze, was especially affected. The city had seen much work designed to build many sponge-city features, but the unfinished project was no match for the flood waters. More than 165,000 hectares of cropland in Sichuan were damaged by the flood, while an area home to two hundred thousand people had to be evacuated. In all, more than two hundred people lost their lives during the flooding episodes. Nevertheless, the death rate from flooding that year was less than half the average figure for each of the past five years. A better warning system reduced fatalities, and Shanghai itself was largely spared serious damage due to flood water. But in order to accomplish this, the hinterlands were "put on the frontline," as one observer put it.[18]

When considering what might happen in the future, it's important to note that the 2020 floods around Shanghai were caused more by the volume of rainwater coursing down rivers than by storm surges and attacks from the sea. Floods after storms will become increasingly dangerous, as increases in rainfall and more extreme weather are part and parcel of climate change. The summer of 2021 saw massive, sudden floods in Henan Province on the Yellow River. These floods, meteorologists say, were also caused by shifting, violent weather patterns. Battling the cause of this change will require action on another level, and China's role will be very important.

The truth is that over the last twenty-five years the country has become the world's biggest emitter of CO_2 and other greenhouse gases. China has built new housing for hundreds of millions of people, rerouted rivers, shot road and train lines across hundreds of kilometres, and become the factory for the world. This has brought a level of prosperity to the Chinese people that was only dreamed of when the Chinese economy opened up in the 1990s. The beggar at the gate to young Jim's house back in 1941 would not believe his eyes were he around to witness what has been wrought.

The tremendous effort, though, has had deleterious effects on the whole world. The big polluters of North America and Europe — where the per person greenhouse gas imprint is still much larger than that of the average Chinese person — are stumbling toward the goals they've set for themselves, and China has announced some interesting targets for its own carbon reduction. But it's hard, and maybe unfair, to expect China to shortchange its people, to sacrifice prosperity for the climate goals that are not immediately visible.

What is clear is that China and Shanghai now have so much invested in their seacoast development that they are going to be extremely reluctant to let it go. People in the hinterland may be sacrificed — like those who will be left behind in Jakarta when the capital is moved — but the Chinese are going to put up a terrific fight against being submerged by rising seas and floods in order to keep the material progress they've worked so hard to attain.

This is not the situation in the region we'll discuss next, Dhaka and the Sundarbans in Bangladesh. There, retreat from rising water may be the only solution, but the people of Shanghai — and of China as a whole — have passed from a time when disaster and loss were expected to one in which people have much to lose and are likely to fight to keep what they have.

6

DHAKA AND THE SUNDARBANS

Writer and activist Amitav Ghosh begins his novel *The Hungry Tide* by telling how one ancient story evokes the Ganges River as "a heavenly braid … an immense rope of water, unfurling through a wide and thirsty plain." But, Ghosh writes, there is more, which is rarely recounted: The braid is a river that "comes undone" to become "hundreds, maybe thousands of tangled strands." It is both the treasure and the despair of Bengal. "There are no borders" as the river nears its end at the Bay of Bengal, nothing "to divide fresh water from salt, river from sea. The tide reaches as far as two hundred miles inland so that thousands of acres of forest disappear underwater only to re-emerge hours later."

Ghosh writes, "The currents are so powerful as to reshape the islands almost daily — some days the water tears away entire promontories and peninsulas; at other times it throws up new shelves and sandbanks where there were none before."

The region is called the Sundarbans, which means "beautiful forest" in Bangla and in Bengali, as the language is known in India. Ghosh, who was born in the Indian state of West Bengal and has family connections with the independent country of Bangladesh, has written two novels set in the Sundarbans. *The Hungry Tide*, in particular, contains some of the most affecting writing about weather I've come across. It is also a chilling analysis of what was happening at the turn of the twenty-first century in the flood-plain and on the myriad islands of the river delta.

The Ganges, in fact, is only one of three great rivers that come together before the delta, and in Bangladesh it even has another name, the Padma. It rises in the Himalayas and then runs south and east across the great, fertile Gangeatic Plain. There it is joined by the Brahmaputra-Jamuna, a river whose origins are on the other side of the Himalayas and which runs eastward along the cordillera before turning west and south across part of India and into what is now Bangladesh. Farther south, the Meghna joins the combined flow. By some accounts the delta formed by the combined rivers rolling into the Bay of Bengal is the biggest in the world. Certainly, it is composed of a multitude of channels and streams running around islands that shape-shift like a wizard in a mythic tale. So much water is carried into the ocean by the river system and other, less impressive rivers that the waters of the Bay of Bengal are the least salty of any ocean area.[1]

The coastal region, the Sundarbans, is shared now by India and Bangladesh. The portions in each country have been designated as national parks, include wildlife sanctuaries, and are also protected as UNESCO World Heritage Sites. The region is, says UNESCO, a "complex network of tidal waterways, mudflats and small islands of salt-tolerant mangrove forests ... known for its wide range of fauna, including 260 bird species, the Bengal tiger and other threatened species such as the estuarine crocodile and the Indian python."[2]

The heroine of *The Hungry Tide* is an American biologist of Indian descent who is studying the freshwater dolphin that is native to the Sundarbans. Her adventures illuminate the problems of this peculiar mix of land and water that is under pressure from increasing population, threatened by industrial pollution, and menaced by rising sea levels.

Most years, the river systems flood part of the region, carrying with them billions of tonnes of fertile sediment, which, when deposited, replenish the land just as the tides do on the Bay of Fundy and the Kamouraska coast of the St. Lawrence, making for rich farmland. But extreme weather is increasingly upsetting the balance between a good flood and a bad one.

The same systems of monsoon rains that inundated China in early summer 2020 caused flooding in the Ganges-Brahmaputra delta that affected 5.4 million people and caused the death of nearly two hundred, 70 percent of whom were children.[3] Part of the high water was due to storm surges and locally heavy rains, but the flooding was also the result of dams upstream in India and China letting out water to avoid overflowing as the rains came down. Because water released from dams is both more abundant and faster moving than the usual volume of streams, erosion was great. The damage was compounded because the previous four years had also been years of big floods, and many dikes and other water-management structures had not yet been repaired.

Much of that infrastructure had been well-thought-out and built with an eye to preventing just what happened, but some of those "good" ideas are turning out to have important downsides. For example, embankments to protect urban areas, as have been built in Dhaka, the nation's capital and biggest city, have frequently presented their own problems. They may work — until they don't. That is, the banks keep the water in check up to a point, but if there is a breach, the pent-up water floods widely as it escapes.

That happened in July 2020 in Dhaka, even though the city lies a long way from the Bay of Bengal. The mangroves and sandbars of the Sundarbans provided some protection — Bangladeshi officials say that the Beautiful Forest saved the country as a whole during Cyclone Amphan earlier in 2020[4] — but they cannot fully protect the city from the double whammy of lots of rain and rising seas.

The problem is compounded by the fact that the trees in the Beautiful Forest — and elsewhere — are under attack. This is nothing new, really; humans have been chopping down forests since before we settled down to grow things. These days, though, the stakes are higher than ever, and the pace is even faster.

Eons ago, before the discovery of how to make metal tools, progress against the forest, any forest, was difficult because hacking away with stone tools takes a long time. Therefore, people turned to fire to clear territory. Tales of huge forest fires show up in folklore, and even in one of the Hindu faith's major texts. Dating back to around the ninth century BCE but compiled later, the Mahabharata tells of the burning of the Khandava Forest on the southern edge of the Gangeatic Plain, near modern-day New Delhi. To look at the countryside around the sprawling, densely populated Indian capital, you'd never think the banks of the Yamuna River were ever wooded, but according to the holy book, the gods Krishna and Arjuna once picnicked in the woods by the river. They were comfortably installed when a poor Brahman came by begging. They gave him alms, but as in so many stories of the magical and religious, the monk was not what he appeared to be. He revealed himself as Agni, the god of fire. No piddling alms would satisfy his hunger. What he needed was to consume the whole forest and all the creatures who lived there.

The pair agreed to this (well, does anyone ever argue with a god as powerful as the one of fire?), and so Agni gave them a chariot and bows and arrows to shoot fire into the heart of the woods. But setting fire to the forest was not enough. Agni told them to drive back all who tried to escape the fire until both the trees and the creatures who live among them were consumed.

Some anthropologists say this is a metaphorical description of what happened when a sophisticated, urbanized people living in the Indus Valley, where they had established towns and farms, moved onto the forested plain, pushing aside the hunters and gatherers who had roamed the territory for perhaps thousands of years. Here again, it seems that old stories record what really happened, because there is evidence that actual fires were set again and again until the newcomers held sway over the land. Contemporary cultural, genetic, and linguistic evidence also suggests strongly that the people who live there now are descended from those who brought agriculture to those forests, just as the ancient Hindu text suggests.

In Bengal, farther to the east, the forest was also the "natural state before humans attacked it with ax and plow."[5] The shifting river courses would sometimes undermine parts of the land, but the sediment deposited by the rivers created new land where more forest would grow. The delta had been inhabited by as early as the eleventh century BCE by people who had a well-developed culture, living in aligned houses, burying their dead in cemeteries, and using weapons and tools made of iron. They probably weren't sedentary farmers; instead, it's likely that they were people who used a slash-and-burn approach to growing food. That is, they would shift cultivation after a few years, clearing new land with fires so they could plant dry rice and millet. Forest would return to the fallow land, and the ecological balance was more or less maintained.

By the sixth century BCE, however, settled farming that required the permanent clearing of forests took over. Part of this change was due to the influx of the Aryans, those people from the north and west who brought with them new languages and skills, as well as religious practices. Because the Ganges was considered sacred by these newcomers, they were particularly attracted to the western part of the delta, where the Ganges was dominant. Later, the river's main channel shifted eastward so that it met the Padma River before it rolled on to the sea. The combined river might have carried a larger load of sediment, but to some extent its spiritual and symbolic load was lessened.[6] The result was, among other things, that the people and their culture on the eastern side of the delta were in many aspects different from those on the western side, which was linked more closely with the Aryan heartland. Historian Richard M. Eaton argues that this difference underlies the subsequent history of the areas that are now the state of West Bengal in India and the nation of Bangladesh.

During the years of the first wave of Islamic expansion, the eastern part was a more fertile ground for Islam since the ties with Hinduism and Aryan culture among the people who lived there were fewer. The first contact with Islam was very early in the history of the religion: the first completely authenticated reference to Muslims in Bengal dates to the early ninth century, less than two hundred years after the death of the Prophet Mohammed, when an Arab geographer Al-Mas'udi noted their presence there. Five hundred years later, Islam entered South Asia again when Muslim Mughal invaders left Afghanistan to sweep south and west, with Bengal at the end of their triumphant conquest. The brand of Islam they brought with them was different from that which had been in place already.[7] Eaton says that rather than being something foreign, imported by the Mughal invaders, Islam in Bengal was already strong, the belief system of the ordinary folk, and — this is extremely important — "a religion of the plow."

He cites an epic poem composed by Saiyid Sultan that says that Adam, the father of the human race, made his earthly appearance on Sandwip Island, off Bengal's southeastern coast: "There, the angel Gabriel instructed him to go to Arabia, where at Mecca he would construct the original Ka'aba," which became the most sacred site in Islam. "When this was accomplished," Eaton writes, "Gabriel gave Adam a plow, a yoke, two bulls, and seed, addressing him with the words, 'Niranjan [God] has commanded that agriculture will be your destiny (bhāl).'" So Adam "planted the seeds, harvested the crop, ground the grain, and made bread," and, says Eaton, "Muslim cultivators still attach a [great] significance to Adam's career" as a farmer. They "identify the earth's soil, from which Adam was made, as the source of Adam's power," and believe that mankind's fundamental task is "farming the earth successfully." This philosophy set the stage in eastern Bengal for "unparalleled growth" of agriculture "as vast stretches of forest were cut and its land cleared for cultivation."[8]

One of the things cultivated was cotton. There were several sorts, but none as prized as *Gossypium arboreum* var. *neglecta*, a shrub-like variety growing as tall as two metres and yielding a fibre quite different from that of other cotton plants. Called *phuti karpas* locally, it was grown in a rather limited area about twenty kilometres southwest of Dhaka along the banks of the Meghna. Seawater, one British colonial official wrote in a citation that sounds like a description of the lush meadows created behind *aboiteaux*, was responsible in part for the variety's special properties. That was because the seawater, "mixing, as the tide rolls it in, with the water of the Megna which overflows that part of the country during three months in the year, deposits, as it subsides, sand and saline particles which very considerably improve and fertilize the soil."[9]

The land could produce two crops a year. Each flower produced a cotton boll that contained between twenty-seven and forty-five cotton seeds surrounded by fluffy filaments that were removed by hand. Then the fibres were spun into thread, mostly by Hindu women whose skill, said one observer, rivalled the handiwork of Arachne, the storied weaver of Greek mythology.

What they produced was remarkable. Unlike other sorts of cotton, this variety's fibres became much stronger when soaked in water, so that ultra-thin thread could be spun from it. Cloth woven from the thread — called *muslin* — could be so fine that it was transparent. The Roman author Petronius called it "woven wind" in his *Satyricon*, adding, "Thy bride might as well clothe herself with a garment of the wind as stand forth publicly naked under her clouds of muslin." Other cultures compared cloth woven from the fibres to "evening dew" and "flowing water."

Dhaka — or as it was called originally, Dacca — was the centre of the industry, an important centre for the Mughal Empire, and a transshipment point for textiles and other wealth. Located a good two hundred kilometres from the open waters of the Bay of Bengal, at first it was a small rural settlement on ground somewhat higher than the plain, which meant that it was surrounded by swamps and marshes.

The Mughal Empire was fabulously wealthy, and multiple images survive of Mughal dignitaries wearing diaphanous skirts of muslin. One dating from 1665 shows two greeting each other with their skirts flaring out like tutus and their soft pyjama-like pants showing underneath. We know that the men are important because one carries a ceremonial fly whisk, which was a symbol of authority all across South Asia. Other royals prized the fabric too; Marie-Antoinette, wife of Louis XVI and the last queen of France, was often painted wearing muslin dresses.

But the portraits were painted as the centuries of fine cotton production were ending. Even after the demise of the Mughal Empire, Dhaka bustled with life and luxury up until the turn of the nineteenth century; in 1800, British officials estimated that it had two hundred thousand inhabitants.[10] This prosperity rather quickly disappeared after British commercial interests effectively destroyed the industry with tariffs, tyrannical destruction of the tools needed to produce the cloth, and ultimately, massive imports of cheaper machine-made thread and cloth produced in Britain.[11]

By 1838 Dhaka was home to only 68,038 people. James Taylor, writing at that time, says, "A great number of houses are unoccupied or in a state of ruin. Drains, ghauts, lanes, and bridges are neglected from the want of funds to keep them in repair. The suburbs are overrun with jungle, while the interior of the town is filled with stagnant canals and sinks, containing refuse animal and vegetable matters, which taint the water of the neighbouring wells."[12]

So total was the dismantlement of the Bengal industry that even the particular variety of cotton on which the whole process was based effectively went extinct. By the turn of the twenty-first century, the only samples extant were botanical displays kept in a handful of herbariums attached to academic institutions. In 2018 a multi-university team announced that they had unravelled the samples' DNA and suggested that comparison with wild varieties might lead to the resurrection of the plant, and possibly of the fine muslin industry in Bangladesh — but that is getting ahead of the story.[13]

For several decades after the collapse of the textile industry, Dhaka's fortunes were linked to a few other industries, like the trade in jute, and to government and other offices set up to manage the region. The city became an administrative centre, with a relatively large number of colonial employees, some of whom had big ideas for the town. For example, in the 1850s, Divisional Commissioner C.E. Buckland ordered the construction of an embankment along the waterfront of the Buriganga to hold back river waters. Called a *bund*, from the Hindi word for dike, it was not an original initiative; the famous Shanghai Bund was inspired by the same movement to protect waterfront property, while the Embankment along the Thames in London was constructed about the same time. The Buckland Bund was created not just for flood control, however. It was also designed to be "a picturesque promenade" where people could stroll and enjoy "breezes off the river," to repeat phrases found frequently in descriptions of the period. Several wealthy residents built imposing houses with large gardens not far away, setting the tone for a Dhaka that was neither the centre of a unique textile industry nor the decaying remains of a town left behind by the colonial capitalists who had destroyed that industry. Note that at this time most of the transport in Bengal was by water, and so the waterfront on the Buriganga — the main river on which the first settlement was built — was the gateway to Dhaka. Only when the first railway was built some distance from the river did the centre of commercial activity shift to the east. The riverbank gradually lost its prestige, with the waterfront devoted to the comings and goings of small freighters, ferries, and other small craft.[14]

Given the shallow draft of the river at Dhaka today, most exports must be shipped by train, truck, or riverboat from the industrial hub to Bangladesh's deep-water ports. The biggest of these is Chattogram, a port mentioned by historians and geographers of the West going all the way back to Ptolemy.[15] Recently, a deal has been signed with Japanese interests to develop a larger deep-water port in nearby Matarbari.

But the needs of twenty-first-century trade were far in the future when the Scottish town planner Sir Patrick Geddes came up with a carefully considered plan for Dhaka in 1917. It was one of several Geddes produced for British colonial cities between 1915 and 1919, but it clearly wasn't just another iteration of one-size-fits-all town planning. Geddes took particular care in his extensive report to consider the physical model of the Bengal delta and the problems and advantages that a landscape filled with canals, ponds, streams, and marshlands offered for an urban environment.[16] He called for a few wide roads, but also for the maintenance and improvement of a network of canals that would connect neighbourhoods and markets. The disgustingly unsanitary state of many of the waterways and ponds and the lack of a sewage collection and treatment facilities received particular attention, as did the health hazards of swamps. But little came of his suggestions. Indeed, the first comprehensive plan for the city was not formulated until the 1950s, in the aftermath of the spectacular population movement that followed the partition of the subcontinent into two countries, India and Pakistan, the latter itself divided into an eastern and a western portion.

The blueprint for the two new nations was drawn up in London and put into effect in 1947, with many consequences that were both unexpected and tragic. Thousands, if not millions, of people were displaced in one of the biggest non-war-related civilian migrations in history. Muslims left India and Hindus left Pakistan, and the great Bangla-speaking region of the Ganges-Brahmaputra river delta was split in two. The population of Dhaka grew from 239,000 in 1941 to 336,000 in 1951, but growth really took off in 1952, the year before Pakistan declared Urdu the new nation's official language. This move relegated Bangla to the sidelines, and the reaction in what was Bangla-speaking East Pakistan was immediate. Two decades of unrest and conflict followed — locally called the War of Independence. This ended finally when the Bengal portion fought itself free of Pakistan in 1971 to become Bangladesh, where Bangla is the official language.

Fifty years later, population estimates for Dhaka run from ten to fourteen million, while the country's population is more than 166 million. A measure of the unbridled growth is the fate of the famous Buckland Bund: it no longer is a place for strolls because it has been built upon, both legally and illegally, and now is crowded with shops, open markets, and other buildings. It is, reports the Dhaka *Daily Star*, "one of the most congested and dirty roads in the capital. Endless streams of rickshaws and goods-carrying vans create traffic gridlocks all day long."[17]

So many people — where to put them? Megacities around the world have been faced with these problems, expanding as far as the eye can see in many places. The issue is particularly acute in Dhaka and more generally in Bangladesh, where rivers have forever reworked the landscape, taking away in one place and building up in others. One example can be seen in comparisons of old maps of Dhaka. When Patrick Geddes was making his plan, a large island in the Burigana stood just to the west of the city. He wanted to plant it with trees and bushes in order to make a public greenscape, but the island disappeared not long after, when the river changed course.[18] Similarly, in Jeremy Seabrook's book about the textile industry in Bangladesh and England, *The Song of the Shirt: The High Price of Cheap Garments, from Blackburn to Bangladesh*, workers he interviews tell him again and again that the disappearance of their homestead and fields into raging river waters is the reason they headed for Dhaka in recent decades.

The excitement of the big city is an attraction, too, of course, but what these migrants find is far from paradise. The picture Seabrook paints of the city in the early twenty-first century is bleak: too many people, working too hard for too little.

> It is not only the fragile land that is eroded by restless rivers and tides; ways of life, cultures, and traditions are also washed away. Even the people are thin, two-dimensional emblems of poverty and the subject of reports, abstracts and inquiries which have been stored away in monsoon-stained files half eaten by white ants, or now, absorbed into seldom accessed cyber data, lost in the info-glut of

modernity. Microcredit initiatives, aid programs, and development projects have come and gone, but the poor remain: bony rickshaw drivers, emaciated elderly maid-servants, children breaking bricks in desolate yards, faces powdered with red dust, while in the drenched villages they lure birds and fish into bamboo traps, a work increasingly necessary [in land] which cannot provide sustenance for its people.[19]

But there is one bright spot, Seabrook says: the thousands of colourfully clad young women who are now working in Dhaka's garment factories. Their lives have been transformed:

> [They] no longer have to be agricultural labourers, maid servants or sex workers if they want to enter the labour force. Young women now have a voice in their family and society. They cannot go back to their former lives of submission and domestic drudgery. In the beginning, religious leaders in villages issued fatwas against women going into factories; but so many people now depend upon their remittances that the religious authorities no longer have the same power over those working women whose income plays an important role in the survival of whole communities.... This is a positive transformation that cannot easily be reversed.[20]

The former *New York Times* columnist and reporter Nicholas Kristof would agree. In a 1991 article, "The Everyday Disaster of Life in Bangladesh," he took a careful look at the country in the aftermath of a cyclone in which perhaps one hundred thousand people had been killed. The country, he wrote, in some respects at that point resembled "countries like China before their Communist revolutions. Not only is it poor — per capita income is $170 a year — but social inequities are overwhelming, with homeless beggars sharing streets with tycoons riding in imported cars. There is a substantial class of peasants without property, who must till the soil of others."[21]

In March 2021, when he visited the country on its fiftieth anniversary, Kristof found things much different. All his pessimism "was dead wrong, for Bangladesh has since enjoyed three decades of extraordinary progress." And what was the country's secret? "It was education and girls." He goes on to say that education empowered these girls, who "became pillars of Bangladesh's economy." In addition, the country's garment industry, which is now the world's largest exporter of garments after that of China, has "given women better opportunities." Kristof adds, "Bangladesh hasn't had great political leaders. But its investments in human capital created a dynamism that we can all learn from."[22]

Before Covid-19, most indicators suggested that Bangladesh was definitely on the way to — well, if not prosperity, then a considerable improvement in the standard of living when compared to that of a few decades ago. In March 2021, the World Bank noted that poverty had declined from 44 percent in 1991 to 15 percent in 2016 and predicted that Bangladesh was slated to "graduate from the UN's Least Developed Countries (LDC) list in 2026." Not bad at all for a country that was the world's second poorest in 1971 at the end of the War of Independence.

Observers other than Kristof also point to the success of the garment industry that employs four million people as the cornerstone of this success. The fact that many people have gone abroad to work and send home money also helps fuel the economy. In the plague year of 2020, twenty-two billion dollars were sent home by workers abroad, the equivalent of 6.6 percent of the nation's gross domestic product (GDP).[23]

But in the background loom rising sea levels. The elaborate dance of waters in the Bengal delta that has been going on for years has intensified as more extreme weather events and rising oceans due to climate change are becoming commonplace. Not only are more and more fields and communities being swept away, but also many of those that so far haven't been are affected by salt water infiltrating their groundwater supply.

A documentary on life on Bangladesh's rivers made by the public broadcaster of Singapore lingers on what happens when people are forced to use brackish water for everything. During the monsoon season, tube wells may deliver decent water, but during the dry season, the water becomes undrinkable because of salt water infiltration, and people must buy fresh water for cooking and drinking. At twenty-five U.S. cents for twenty litres, bottled water is a big expense for people who are practising subsistence farming supplemented, perhaps, by remittances from family members working in Dhaka or abroad.[24] The bottled water is far too precious to use for washing or cleaning. As a result, says a woman interviewed in the film, during the dry season they wash with brackish river water and their skin itches and becomes irritated. At first hearing, that doesn't sound too bad, but then the focus shifts to a clinic where the doctor explains that skin ulcers are very common in the area due to constant contact with salt water.

The burden of collecting water from pump, rivers, or stores falls on women, a government official comments, giving another reason why young women might want to get out of villages to work in Dhaka as quickly as possible.

How to attack these problems? The answers are both complicated and simple but putting them into effect is probably easier now in Bangladesh than it would have been twenty years ago. During the last decade or so, while the eyes of the world looked in amazement at the progress being made in Bangladesh in terms of GDP increases and basic support for things like better maternal health, a number of outside interests have proposed ways to improve the situation even more.

The Dutch, probably not surprisingly given their long experience with rivers, deltas, and reclaiming land, have been involved in several endeavours. Shortly after Bangladesh became an independent country, a project with much input from the Netherlands saw the construction along the coast of 139 polders of the sort the Dutch perfected.[25] Using a system of dikes and sluices rather like the *aboiteaux* of French Canada, the polders allow excess fresh water to flow into the sea at low tide, leaving behind fertile sediments. These are managed in large part by the farmers who live around them, and it is hoped that the knowledge they acquire here can be transferred to other areas.

These efforts, coupled with a vastly improved early warning system, have reduced mortality from flooding remarkably, although the polder system is not without its problems. In 2009 one polder established in the 1960s famously turned into a small-scale disaster when the embankments protecting about eighty square kilometres of villages and fields were breached.[26] Because the land inside had not been replenished with sediment over the years, it was much lower than the usual level of waters surrounding it, and salty water flooded in as if into a bathtub. The result was hard times for the people who lived there, but it prompted reflection on how to manage the situation. Periodically allowing sediment-bearing water to enter a polder is now considered good polder management by some. Certainly, keeping the myriad channels open and policing erosion on the embankments require a lot of effort, and to date only a relatively small number of people have been involved in the ongoing work — around sixty thousand. However, the effort protects a far larger population from flooding and mounting salinity: about nine million. Both the Dutch and the Bangladeshis hope that the lessons of co-operative work learned here can be translated to other places and other people.

In 2018 Bangladesh approved the Bangladesh Delta Plan 2100, which was elaborated with much input from Dutch experts and interests. The idea was to capitalize on what's commonly called the Dutch Delta Approach to come up with a long-term solution "to achieve a water secure, flood safe, climate resilient and prosperous delta, which ensures long-term water and

Bangladesh's fields are frequently flooded by rivers and seawater.

food security, economic growth and environmental sustainability by means of robust, adaptive, integrated planning strategies and equitable water governance."[27] Behind the Dutch initiative is a desire to make a difference in the world and, let's be frank, to sell Dutch know-how and equipment to other countries.

For Bangladesh, the effort fit into the country's long-term and ambitious plans to become a middle-income country. To accomplish this, a number of programs had already been undertaken, including making contraception free and available for all and instituting widespread mother and child health care. The success of this kind of program is evident: Bangladesh has lowered its birth rate from 6.94 children per woman at Independence down to only slightly below the replacement level — 2.003 in 2019.[28] Notably, it did this without forcing sterilization on its people, as India had done at times, or placing a limit on the number of children a couple has, as China did for several decades. The key would appear to be educating girls and demonstrating that a family needn't have many children in order to assure that two or three make it to adulthood — cutting down on the deaths of mothers and babies and eradicating childhood disease showed that this approach to family planning could be successful. As the result of a concerted effort, the mortality of children under five dropped from 244.68 deaths per thousand births in 1971 to 28.95 per thousand in 2020. Among the techniques used was treating childhood diarrhea through a number of tactics, including making packets of rehydration salts available free or for a few pennies everywhere.

Bangladesh's Delta Plan 2100 contains other, far more high-tech elements. In total, there are eighty projects. Sixty-five are physical ones, while fifteen are institutional and knowledge-development ones scheduled to be undertaken during the first phase, which runs up to 2030. The plan's total capital investment cost will be BDT2,978 billion (US$37 billion), with money coming from Bangladesh's own budgets, from international funds, and from private foreign investors.

That's serious money, but other sorts of projects are needed, too. One involves protecting and possibly expanding the extensive mangrove forests that make up the Sundarbans and dampen the fury of rising seas and extreme weather events. We'll talk about them more in chapter 8, "What Can

Be Done — What Will Be Done?," but for the moment just note that Bill Gates thinks they're a big part of the solution to the problem of impending climate disaster. When yet another cyclone pounded Bangladesh in May 2021, coinciding with a major high tide and causing a storm surge of about 2.5 metres, flooding damaged nearly 40 percent of the region's cropland, but devastation would have been much worse had not the mangroves of the Beautiful Forest tempered the storm.[29]

Then there's a startlingly low-tech project: making floating gardens to get around the problems of what happens when the waters rise. Using platforms made of such things as bamboo and water hyacinth plants as a base, farmers can raise a number of vegetables and herbs for their own use or for sale.[30] The method is nothing new — it's been used in some places on the Bengal delta for hundreds of years, but a concerted effort to spread the word about the technique is being made. According to a report by the Food and Agriculture Organization of the United Nations, making one costs about $94, but farmers can make $140 per square metre of garden if they sell their produce.

It's significant that the photos accompanying reports and articles about the floating garden initiative show people up to their knees, if not their waists, in water as they work on their garden rafts, some with smiles on their faces. They contrast markedly with the shots of Dhaka during floods, where rickshaw cycle drivers peddle furiously through streets that look like stream beds. In the first, people seem relatively content with what they're doing. Working in water is something they know, something they are controlling in a small way. The rickshaw driver in a monsoon seems far from happy, however.

While regular flooding and the combination of erosion and formation of sediment-fed islands elsewhere have been a feature of the Bengal delta for thousands of years, climate change means exponential increases in the effects of these cycles. Where the flooded-out will go is a question that is worrying many in the country, as well as elsewhere. As one group of researchers write, "Within South Asia, Bangladesh stands as the most vulnerable: 4.1 million people were displaced as a result of climate disasters in 2019 (2.5% of the population), 13.3 million people could be displaced by climate change by 2050, and 18% of its coastland will remain inundated by 2080."[31]

Dhaka itself, even though it is a couple of hundred kilometres away from the Bay of Bengal, will be one of the world's major cities most affected by rising seas. At sea levels that will follow an average temperature increase of 1.5°C, all but the higher parts of the city will be submerged. If the increase in temperature is 3°C, even the Lalbagh Fort, now protected by dikes and surrounded by elaborate gardens, will rise out of the water as if from a reflecting pond, according to a simulation by Climate Central.[32]

Some in Bangladesh propose programs that would designate smaller centres around the country as places were people displaced by climate events could relocate. Seven widely scattered towns are currently involved in making migrant-friendly urban areas.[33] These include improvements in sanitation and water supply in areas where the migrants might settle, training programs for the newly arrived, and, in one case, community-organized, low-cost, but climate-resilient housing. The authors of the study, all of whom are active on the front lines of relocating migrants, recognize that their efforts won't solve all problems and that some international migration is inevitable. That is nothing new: migrants from Bangladesh have been leaving the country for years.

Hearing Bangla everywhere he went in Venice was a major inspiration for Amitav Ghosh's more recent novel *Gun Island*, which is a sort of sequel to *The Hungry Tide*. He says that he was struck by the numbers of migrants from Bangladesh, legal and illegal, who were working on gondolas, making pizza, and repairing water-damaged buildings in the Italian tourist mecca. The dialect they spoke was the same one he had used with his grandparents before he left South Asia decades ago. The reasons why they were now in Venice are an integral part of his novel.[34]

But all of these things are strategies, not solutions to the problems of climate change and rising sea levels, which are going to get worse. Getting to the root of the problem will require far more effort than either the Dutch or the Bangladeshis can produce. It is going to take united, immediate, worldwide effort to cut down on greenhouse gas emissions to keep global warming to under 1.5°C. Bangladesh's government says it is doing its part, but the "developed" world hasn't been. As Prime Minister Sheikh Hasina said in her address to the COP26 meetings in 2021,

Bangladesh has one of the world's most extensive domestic solar energy programs. We hope to have 40% of our energy from renewable sources by 2041. We have cancelled 10 coal-based power plants worth 12 billion dollars of foreign investment. We are going to implement the 'Mujib Climate Prosperity Plan' — a journey from climate vulnerability to resilience to climate prosperity. We are trying to address the challenge of climate impact because of 1.1 million forcibly displaced Myanmar nationals or Rohingyas.

As the Chair of the Climate Vulnerable Forum (CVF) and V20 (Vulnerable 20), we are promoting the interests of the 48 climate-vulnerable countries. We are also sharing best practices and adaptation knowledge regionally through the South Asia Office of the Global Center of Adaptation's Dhaka. On behalf of the CVF, Bangladesh is pursuing to establish a Climate Emergency Pact.

She ended with four direct demands to the major emitters of carbon dioxide. First, they must raise the standards for what they will do at home and implement these plans. Second, "developed countries should fulfil their commitments of providing 100 billion dollars annually with a 50:50 balance between adaptation and mitigation." They should also "disseminate clean and green technology at affordable costs to the most vulnerable countries. The development needs of the CVF countries also need to be considered." And last, "the issue of loss and damage must be addressed, including global sharing of responsibility for climate migrants displaced by sea-level rise, salinity increase, river erosion, floods, and draughts."

Will this happen? We'll look at that more closely in chapter 8, but for now note that Covid-19 has made everything that happens in Bangladesh that much harder.

Nevertheless, if I were a betting person, I'd put my money on Bangladesh to ride out this wave, the way its people have been riding out the floods for generations, even though the flooding is getting worse. The same kind of social programs that have resulted in such dramatic changes in birth rates,

infant mortality, and educational attainment since the early 1970s may —
just may — see the country through to a kind of accommodation with climate change.

7

THE SALISH SEA, VANCOUVER, AND SEATTLE

S hortly before the *Dhaka Tribune* reported in mid-2021 on surging Covid-19 cases in Bangladesh, news media in North America and around the world almost incredulously reported on a heat dome that had settled at the end of June over the Pacific Northwest and parts of western Canada. Temperatures soared higher than ever before recorded in a region famous for rain and cool weather. But on June 29, according to the *Washington Post*, it was literally hot enough to cook seafood. Alyssa Gehman, a marine ecologist who lives near the Salish Sea in Vancouver, B.C., went to the beach in an attempt to beat the heat, the newspaper reported. It "was packed with others," which she expected. What she didn't expect was the smell: of "putrid shellfish — cooking," she told the *Post*. "All around her, beds of mussels had popped open, dead," because of "the heat beating down on the rocks." In all, "an estimated 1 billion small sea creatures" along the more than 6,400 kilometres of shore around the Salish Sea perished during the heat wave.[1]

One of my first memories is of the Salish Sea, only it wasn't called that then and it certainly wasn't that hot, so many years ago. It was during the Second World War; my father was in the U.S. Army, stationed at Fort Lewis, south of Tacoma at the end of Puget Sound. Because of the wartime housing shortage, we were living in a big old Victorian house that had been broken up into minuscule apartments — ours was essentially one big room with a hot plate and sink in one corner and the bathroom down the hall. Even if my dad was away a good deal of the time at the base, it was pretty crowded. Not surprisingly, then, when the weather began to veer toward spring, we went outside. One Saturday morning, bright and early, my dad and one of his friends set off to Point Defiance Park to fly a kite they'd made.

Did we take the bus? Did someone have a car? Did we live close enough to walk? I have no recollection of that, but what I do remember was the way the kite took off into the air, swirling out across the street and then beyond, toward the blue expanse of the sound. It was, I think, the first time that I looked beyond my immediate surroundings, which was a world that included little above the height of a three-year-old not much taller than the knees of grown-ups. When they got the kite in the air, my father picked me up so I could look over the houses and the trees to where it was sailing over the water, over the Salish Sea.

Since then, I've spent quite a bit of time in the area: parents in Seattle and Bellingham for a while, sister and family in Vancouver, cousins in Bothell. For more than a decade, I drove frequently up I-5 and B.C.'s Highway 99 between the south and the north of the Salish Sea, and even back in the 1980s, I remember worrying about sea-level rise. Would the Leopold — an elegant old hotel turned partly into a residence for retired people in downtown Bellingham, where my parents were living — be affected, seeing as Bellingham Bay was a stone's throw away? As it turned out, there was no reason to fret on that front. It hasn't been, not even in the floods of late 2021. In the future, it probably won't be either, because the original building went up in 1889, back when people had some sense about where to build; that is, up the hill a bit from the waterfront, looking out toward Lummi Island.

But things are going to change around the Salish Sea.

Which is what exactly? Even though the name was officially approved by a raft of governmental bodies in both Washington State and British Columbia more than ten years ago, many people, including some who know the Puget Sound, Strait of Juan de Fuca, and Georgia Strait well, can't immediately place it.[2] Among them is my husband, who also has spent quite a bit of time in the region but needed an explanation when I started talking about it.

Look at a map of the West Coast from the Washington-Oregon border north, and you'll see the bodies of water that stretch from near Olympia, Washington, in the south to the top end of Vancouver Island in the north. The names they have carried for the last two hundred years or so don't reflect the people who originally lived around them; instead, they bear the names of the Europeans who "discovered" them. The Georgia Strait (the waters between Vancouver Island and the mainland of British Columbia as far as, say, Campbell River) was named as a tip of the hat to King George III, the British monarch when Captain George Vancouver made his voyage up the Pacific Coast of North America from 1791 to 1795. Vancouver named Puget Sound, the long bay that extends to the south, for Peter Puget, a second lieutenant in his expedition who probed the main channel of the sound. The sound and Georgia Strait are separated by the Strait of Juan de Fuca, which cuts between Vancouver Island and the Olympic Peninsula. In large part a deep marine canyon, it is the link between the open ocean and what, many scientists say, is in reality an inland sea. The strait bears the name of a Greek navigator who sailed for the Spanish in the seventeenth century and who apparently was the first European captain to see it.

The boundaries between these three bodies of water were never very clear. For example, tell me at what point did Puget Sound become the Strait of Juan de Fuca? And what about those islands in the middle called the San Juan Islands in the United States and the Gulf Islands in Canada? The international boundary between the United States and Canada complicates things. On the mainland, it runs straight along the forty-ninth parallel until it reaches the salt water, but then it swoops through the islands before dividing the narrows between Vancouver Island (in British Columbia) and the Olympic Peninsula (in Washington State). Certainly, the First Nations (in Canada) and Native Americans (in the States) who lived around here didn't divide it up in such arbitrary ways.

Until the 1970s and '80s, questions about what to call the bodies of water were Shakespearean: you know, "a rose by any other name would smell as sweet." But as questions about the transport of oil from the Alaskan oil fields to refineries in Washington State began to be posed, it became clear that larger terms of reference were needed. Research on both sides of the international border began to spotlight the ways in which the three bodies of water were, in fact, an ecosystem that was best understood as an "integrated inland sea."[3] In fact, it is an estuary into which several rivers, including the mighty Fraser, pour their waters and where this fresh water is mixed with salt water flowing in with the tides from the Pacific Ocean through the Strait of Juan de Fuca.

After it was agreed that it was basically one body of water, the question arose of what to call it. Enter Gary Weber, one of the scientists involved in research on the inland sea's ecosystem. He had loved it since childhood and had ties to both the Canadian and the U.S. sides of the border. In the 1990s, he pointed out that all the Indigenous groups living around the inland sea "shared a historical connection with the Coast Salish language[s]." Calling it the Salish Sea recognized their important link and was far less cumbersome than other names proposed, such as the Georgia Basin Puget Sound Ecosystem. Slowly over the next twenty years, the name gained supporters, until by the summer of 2010 it had been recognized by the relevant governmental entities as well as by Indigenous communities on both sides of the border. Today, about eight million people live on its 7,470 kilometres of coastline, which is where our consideration of rising sea levels comes in.

≈

Decoding the relations between sea and land in the Salish Sea is a complicated business. As noted earlier when we spoke about Doggerland and the Storegga landslide, the background story of what happened when icefields melt has two parts. Sure, the added liquid water meant that there was more water in the seas, so that sea levels rose, but as the ice melted, parts of the Earth underneath it rebounded as weight was decreased. This means that in some places areas that were once on the coast have been lifted metres into the air, earthquake by earthquake, as the land rebounded, while in others, where the rebound was less or non-existent, coastlines have been slowly drowned.

A second kind of earthquake frequently has even greater effects than those caused by rebound of the Earth's crust. My own introduction to this force of the Earth came at precisely 7:18 p.m. on February 14, 1946, when a large earthquake occurred between Tacoma and Olympia, Washington. I was about three and in the bathtub of the shared bathroom of the rooming house where we were living. I don't remember the noise or the moving floor and walls, just being bundled up in a towel by my mother and hustled outside to where the other residents were gathering. Nobody was hurt, although it's possible that some bricks fell off the chimney, and in a very short time I was back inside, sleeping in my tiny bed.[4] But I've had a lot of respect for earthquakes ever since. This was the most frightening one that I remember, even though I grew up in Southern California and experienced two larger quakes in Quebec, where the Earth is still rebounding from its ice age load of glacier.

These days, the ground moves around the Salish Sea relatively frequently. The reasons for this are in large part linked to the slow dance of sections of the Earth's crust, which we talked about when setting the stage for a discussion of Indonesia. As it happens, several faults run through and around the Salish Sea, where three plates of the Earth's crust come together. The dense Pacific Plate underneath the ocean is moving northwest, and rotating the smaller Juan de Fuca Plate, pushing it under the lighter North American Plate. The relative movement averages out at about four centimetres a year, or the rate at which one's fingernails grow, but the shifting isn't constant. Rather, the pressure builds up over time — sometimes for decades or even centuries — and then is suddenly released as the plates slip against each other. Depending on the depth of the quake, it can be felt over a wide area (usually the case with the deepest quakes) or in a more localized one. The biggest earthquakes around the Salish Sea are those that occur where the North American Plate and the Pacific Plate shift. The last megaquake was in 1700 and is estimated to have registered 9.0 on the Richter scale. Japanese records show that the tsunami from the quake ravaged the shoreline there.[5]

These movements of the Earth have complicated research into the paths taken by people entering and settling in North America. Local uplift has left at least one site used by early sojourners about one hundred metres above sea level.[6] But since this kind of movement ceased about 6,700 years ago in this part of the world, probably the best places now to look for artifacts and other signs are underwater. This is important because over the last two decades, the Pacific coastline of North America has become a focal point for research into the great population movement from Asia fifteen thousand or more years ago.

A quick review for those of you who took Anthro 101 many years ago: About 1930, a discovery of what was thought to be the earliest evidence of humans in North America was made on the windswept plains of New Mexico.[7] The people who lived there were master toolmakers who used a distinctive technique that flaked away chips along both edges of a spear point. It came to be called a Clovis point, after the place where they were first found. By extension, the people who made them were also called Clovis. At the time of the first discoveries, it was assumed that the artifacts dated to about 3000 BCE, but when radioactive carbon dating was developed in the 1950s, scientists found to their amazement that the more than eighty sites where Clovis points had been found dated between 11,500 and 10,900 years ago, when much of North America was covered by ice sheets.

But how did the Clovis people get around the ice blocking their way? They must have travelled through an ice-free corridor between the big ice sheets, many researchers postulated.

Then the story of humans in the Western Hemisphere was pushed back even further. In the 1970s, excavations in Chile uncovered remnants of a campsite with tools, ruins of a wooden structure, and hearths. The footprint of a child rounded out the find, although no skeletons were uncovered. Radioactive dating put the time somewhere around 12,000 BCE.[8]

The finds stirred up a storm of controversy, with some anthropologists angrily questioning the research protocols used. But the underlying problem of how people could have gotten so far south so quickly posed even bigger questions for some researchers.

Other finds followed, which showed both very early spread of humans and a wide geographic distribution. One of the most striking we mentioned before: human coprolites — ancient, dried poop — from a cave in Oregon.

They have been dated to more than 12,000 BCE, a time before agriculture anywhere, before anyone in the world had settled down in permanent settlements, and like the Chilean find, before the Clovis people.

True, the famous 2,500-kilometre-long ice-free corridor — first postulated without much evidence in the 1930s — has indeed existed at least twice, once before the Last Glacial Maximum and then after the massive ice sheets began to melt. But during the window of time that would allow for the Clovis people's migration, the great Laurentide Ice Sheet had coalesced with the western glaciers to make a continuous field of ice.[9] It looks like the way was blocked by ice from about 19,000 BCE until around 11,000 BCE. After that, many of the valleys would have been filled by meltwater lakes. This means that for thousands of years it would have difficult, if not impossible, for early North Americans to use the corridor as a highway.

A recent find in New Mexico of human footprints dating to about 21,000 BCE suggests there is a possibility that an early group of people travelled south before the great icefields formed,[10] but so far there seems to be no link between them and the current Native American/First Nations populations of North America. Perhaps they left no progeny, just as those people who left Africa long before the Great Expansion and whose remains were found in Israel appear to have done. No one knows.

However, there was another way for people like us to settle the Western Hemisphere during the waning days of the Last Glacial Maximum: it was a path that people might have taken from cove to cove and headland to headland down the west coast of North America. The country to the east might still have been covered by thick ice, but the coast would have provided good hunting and fishing, and during warmer seasons, berries and tubers. For some of these people, the places became home, giving lie to the idea that they were rushing south along some predetermined track. The diverse, dynamic coastal environment inhabited by these resourceful people was not primarily a corridor, or a highway or a route, any more than Beringia, the land around the Bering Strait was a bridge and not a subcontinent. The ancient coast of British Columbia was, above all, a place to be, a homeland of the senses for those who were not on the way elsewhere, least of all to southern Argentina.

Sounds quite a bit like what researchers looking at Doggerland have concluded, doesn't it? Furthermore, like the Doggerlanders, for several thousand years these people lived with rising sea levels as the climate became warmer. On Haida Gwaii, where much research has been done, the sea-level change was "on the order of 5 cm/year for two millennia: the camp in which a person was born would be in the lower intertidal zone when they died; their grandparents' camps would be fishing spots or reefs.… Accepting that sea level change was an integral part of their world, not an external imposition … changes the frame from *coping* to *dwelling*."[11]

A lot of the evidence for this is submerged along the coasts of North and South America, and — perhaps especially — in the Bering Sea. As the great ice sheets melted, something like ten million square kilometres of coastal plains were submerged from Beringia to southern Chile. In stretches of the Pacific Coast where the drop-off is abrupt, this has meant that the waves now beat a kilometre inland from where they crashed ten thousand years ago, while in the shallow northern stretches, the sea pushed as much as five hundred kilometres inland.[12] But that has not kept scientists from looking for signs of coastal people and their settlements.

The earliest traces of humans in North America were discovered in the Yukon in 1976 at the Bluefish Caves. Now dated to sometime around 24,000 BCE, they included microblades, burins, mammal bones that appeared to be cut by tools, and flakes and chips that appeared to be the result of humans' making tools. When the finds were reported, questions arose about the dating and whether, in fact, the objects were made by humans or were simply rocks shaped by natural forces like erosion or freeze-and-thaw cracking. Recently, however, more stringent techniques have been used that corroborate the early date. This evidence suggests that people were inhabiting Beringia long before the peak of the Last Glacial Maximum and would have had plenty of time to venture south, way back then or later when the weather began to improve.

Travel down the kelp highway was possible beginning about 14,000 BCE. Some of the earliest findings show up on the British Columbia coast north of the Salish Sea. Here, because of the complicated interplay between rising seas and glacial rebound, some sites formerly at sea level are now much higher, making research easier to conduct. Sites in the Haida Gwaii and on the central B.C. coast have been dated to 12,500 BCE. The absence of evidence for early sites around the Salish Sea itself may be due to the luck of the draw or may reflect the fact that much of the current landscape of islands and open water at the northern end was probably dry land until rising sea levels drowned the lowlands there. This would mean that it would have been necessary to travel down the western coast of Vancouver Island to enter the Salish Sea at the Strait of Juan de Fuca. This long journey may also account in part for the distribution of Indigenous language families in the area, with Salishan languages spoken around most of the Salish Sea and languages from other families spoken to the north, on western Vancouver Island, and on the tip of the Olympic Peninsula.

The oldest evidence so far uncovered of Native American settlements in the southern Salish Sea is at a site near the Sammamish River north of Seattle. It has yielded artifacts dating to 10,000 BCE.[13] There also are hints of what life was like then in stories still told today. One collected at the beginning of the twentieth century depicts a time that sounds very much like the last days of glaciers: It tells of land "locked in ice" where "North Wind sent his freezing blast all over the country." A very unfriendly spirit, "he built an ice dam across the Duwamish River," which now empties into Puget Sound at Seattle. The barrier "stopped salmon from running upstream, so the people were always hungry and cold." This went on for month after month, year after year. But then a love triangle developed between North Wind, a woman named Mountain Beaver, and her husband, Chinook Wind, the warm "snow eater" that sometimes blows down the mountains at the end of winter. According to the story, North Wind killed Chinook Wind. His son, Storm Wind, avenged his father's death by blowing "rain all over the valley" and melting everything. "North Wind ran away. He went farther north … If [he] had not been chased away, we should all be cold and hungry all the time. As it is, we have a little ice and snow, but not for long, only until Storm Wind comes again."[14]

Sounds quite a bit like those stories that First Nations Australians tell about rising sea levels, doesn't it? Dating this one is harder to do, but what is clear is that there were people on the coast a very long time ago.

This idea is corroborated by the discovery of human footprints on Calvert Island on the central B.C. coast. There, the sea was probably about two or three metres lower between 12,000 and 9,000 BCE than it is now, so researchers set out to test in a systematic manner what lies under the current intertidal beach sediments. They found a set of twenty-nine footprints of at least three different sizes, made by people who were living near the shore around 11,000 BCE.[15] Another *wow!* moment was a find on the central B.C. coast on Triquet Island. This one dates to about 12,000 BCE and includes carved wooden tools and the remnants of a hearth. That's impressive in and of itself, but the First Nations people who live there now — the Heiltsuk First Nation — say it simply confirms their oral history. William Housty, a member of Heiltsuk First Nation, told the CBC in 2017, "Heiltsuk oral history talks of a strip of land in that area where the excavation took place. It was a place that never froze during the ice age and it was a place where our ancestors flocked to for survival."[16]

That's not the only story that has gained prominence as researchers dig. Quentin Mackie, an archeologist from the University of Victoria, is sure that an ancient Haida story has its roots in actual climatic events. Back in 2009, he noted that a Haida tale recorded in the 1900s by J.R. Swanton appears to explain why stretches of water between the Haida Gwaii widened over time: the important mythical figure Raven-Walking "pushed the islands apart with his feet" back in an age when "there was no tree to be seen" — that is, when glaciers were melting and forests had not moved northward. Mackie says, rather like Patrick Nunn talking about the tales of First Nations Australians, "The congruence of the geological and oral historical information leads me to conclude that Haida storytelling extends to the terminal Pleistocene and thus spans more than 14,000 years of the history of these islands."[17]

In addition to the work being done in British Columbia, researchers are systematically exploring the submerged landscape off Southern California's northern Channel Islands archipelago. Currently, the island group consists of San Miguel, Santa Rosa, Santa Cruz, and Anacapa islands, but when the sea level was ninety or so metres lower, they formed one island that researchers have dubbed Santarosae. Nearly one hundred sites now above the waterline have been explored, with one containing buried human bones that date to around 11,000 BCE. The enticing possibility that even older sites might be underwater has led to a plan to map the submerged landscape through various techniques.

≈

July 28, 2021

Bulletin

Orcas — sometimes called killer whales — were sighted off the coast of B.C. in the Haro Strait, the first major sighting of the endangered species in more than three months.

Researchers were excited to see the sea mammals because, although the inland waters off Vancouver Island are considered the summer residence area from April to September, until now the orcas have been mostly absent. According to the Orca Behaviour Institute, their absence is connected to a dwindling food supply, probably because of what's happening to Fraser River salmon runs.[18]

≈

The Fraser River is the biggest of the many rivers that flow into the Salish Sea, and at one time, the salmon in it and elsewhere in the Pacific Northwest appeared to be an inexhaustible resource. Six different sorts of salmon as well as steelhead trout hatch in streams sometimes hundreds of kilometres from the sea. The hatchlings descend the watercourses toward the coast, and then pass their adult lives in the ocean before returning to the same stream to spawn. The various types of salmon, as well as their subgroups, travel at different rhythms, which means that for generations the Indigenous peoples could count on an abundant, high-calorie food source nearly year-round.

How those in the Pacific Northwest husbanded this resource for thousands of years has been well-documented. Using a number of fishing techniques, each year they harvested as many fish as were annually fished commercially in the mid-twentieth century. These methods clearly conserved the stock better than twentieth-century practices, however, as salmon stocks have crashed in recent decades because of overfishing and deteriorating habitat.[19] In July 2021 Canada's Department of Fisheries and Oceans closed 60 percent of the B.C. commercial fishery in a move designed to "reset" the situation, but whether it will do much good is unclear, since much of the decline is due to the trashing of the environment where salmon spawn.

The people around the southern part of the Salish Sea were also fishers of salmon, and probably used many of the same techniques to manage the fish that their neighbours to the north did. These include taking particular care of the streams where the fish spawn, making sure that the way is clear for the fish to swim upstream — sometimes even carrying fish around sudden obstacles like rockslides (documented in both 1904 and 1913) and, it seems, occasionally moving eggs and hatchling fry to new watercourses. Fishing was done sometimes from canoes, sometimes with fish traps, and sometimes on the big rivers with harpoons, but nowhere were the fish vacuumed up indiscriminately at sea the way they have been beginning in the early twentieth century. Nigel Haggan and his fellow researchers argue that the approach used by the Indigenous peoples, where individual groups and families had the responsibility to guard the fish that spawned in their territory, should be a model to follow. Certainly, the model being used today isn't working.

This stewardship began to change when folks descended from Europeans started arriving on the West Coast in the mid-nineteenth century. One of the things they set about doing was transforming sloughs and marshes into arable land. As we noted earlier when talking about the wetlands of Doggerland, marshes are underappreciated today, but they have been and are great resources for food and other materials in diverse cultures.

On the many trips I took between Seattle and Vancouver, I was always impressed by the views from Interstate 5 of the wide Skagit Valley, where marshland has been transformed into dry land. The Cascades rose to the east, the river valley stretched out green on either side of the highway, and to the west I could see Fidalgo Island across the water. The river flooded, I knew, but it wasn't until I started working on this project that I began to wonder about the effect the encroaching sea would have on this lush landscape.

One of the impressive things one sees on the drive in the springtime is the display of tulips. As I remembered them, I started thinking, Aha! Those Dutch. They must have been here first and set about diking and protecting this lowland. But no, while the magnificent fields are indeed a legacy of the Dutch, the bulbs were imported from the Netherlands by a Brit at the beginning of the twentieth century and were grown on a small scale at first. The truly spectacular display began much later, when a Netherlander in the area began growing bulbs after the Second World War.[20] The first dikes are much older than that, though, and are directly related to the Canadian *aboiteaux*.

In 1863 Sam Calhoun — who was born in Albert County, New Brunswick, and had followed the westward movement to what was then the Oregon Territory — was working as a ship carpenter at a lumber mill and shipbuilding yard on one of the islands in the Salish Sea near the mouth of the Skagit River. According to a memoir he wrote as a middle-aged man, one day not long after he arrived, he went out in a canoe to explore the marshes on the north side of the river. He writes, "I thought it the most beautiful sight that I had ever beheld. 'Here,' I said to myself, 'is a country within range of my vision that will support a million people. Here is my home where I shall spend the remainder of my life.'"[21]

Calhoun's hometown of Hopewell is on the northwest side of the Bay of Fundy in New Brunswick. Beginning in 1698, Acadian settlers there began building the dike and sluice system that allowed the conversion of tidelands into meadows like those at Grand-Pré in Nova Scotia, across the bay that we talked about earlier. With *Le grand dérangement*, they were forced from the land about one hundred years later, but the system they'd built was taken over and expanded by later immigrants from Germany and the British Isles.[22] Calhoun obviously knew how the *aboiteaux* worked, and he set about building some on the Skagit Valley salt marshes and sloughs. The other settlers in the area thought him foolish. An account of those early days says, "The white men in the other neighborhoods of the sound were very much inclined to ridicule these efforts to make a farm on mudflats, where the tides overflowed, but when the first immense crops were harvested they saw their error."[23]

Aboiteaux require constant upkeep, though, and by the end of the century, Calhoun had retired, and they'd been allowed to deteriorate. Nevertheless, agriculture in the valley flourished with the later adoption of another sort of dike and sluice system in the 1930s. This one involved concrete and steel and is currently the subject of considerable controversy between farmers and the Swinomish Indian Tribal Community. Today, seventy thousand acres of fields on the broad valley floor are cultivated; eighty sorts of crops and seed for a wide variety of vegetables and flowers are grown. However, this water-management system plays havoc with salmon for which the Swinomish have fishing rights.[24]

Currently, negotiations are underway to return up to 2,700 acres of the delta waters to a more salmon-friendly environment, and in September 2021, the Swinomish Indian Tribal Community filed notice that it intended to sue the U.S. Army Corps of Engineers for not moving quickly in restoring the waterways.[25] This would include changing the sort of sluice in order to allow migrating salmon access to the rivers, and planting trees along the water-courses — some of them quite small — to insure that water temperatures do not rise too high for the salmons' successful spawning journey.[26] In the overall picture, these changes will not have a huge impact on survival of salmon throughout the Pacific Northwest, but if repeated again and again, they might.

The conflict between farmers and Indigenous fishers is only part of the larger problem of protecting against rising sea levels, and certainly it is not confined to Washington State. Just north of the border, the extensive agricultural lands of the Fraser River delta are threatened. So are the homes of many of the three million people who live in the Lower Mainland of British Columbia. November and early December 2021 saw atmospheric rivers carry record-breaking rains from the tropics to the region, which resulted in massive flooding along some watercourses. The certainty that extreme weather events like this — and like the heat dome of the previous summer — will reoccur must be considered in any plan to deal with climate change and rising waters.

~

I mentioned way back in the beginning that I've driven the stretch of British Columbia's Highway 99 many, many times. After you cross the border, the highway goes up and down a bit, then drops down into the flatland, which is the floodplain of the Nicomekl and Serpentine Rivers. For a distance, the road runs very close to the Boundary Bay shore, and at one point the rail line that parallels the highway actually crosses part of the embayment on a trestle. The shoreline for most of the way is buttressed by rip-rap and dikes, but generally the tide flats remain to attenuate the surges of tides and waves. The name of part of the area, Mud Bay Park, tells it all — or almost.

When I was travelling along there, I never encountered flooding from either the rivers or tidal surges, but parts of Bellingham were flooded in November 2021, while to the north and east, the Fraser and its tributaries went wild during the torrential rains.

The double whammy of rising sea levels and more rainfall is going to require a lot of work. Even before 2021's disastrous rains, the Province of British Columbia had told municipalities to plan for at least a one-metre rise in sea levels by 2100. To that end, in 2016 the Municipality of Surrey began to plan, in collaboration with residents, community groups, agricultural interests, and the Semiahmoo First Nation, whose home lies on part of the land.

Three years later, after myriad meetings, surveys, and discussions, the Coastal Flood Adaptation Strategy was adopted. It will roll out during coming decades; seventy-six million dollars has been pledged by the Canadian federal government to get things under way.[27] Among the first projects are reinforcement of dikes and sea dams, as well as development of "softer" coastal modifications. These include "living dikes," where sediment will be added to strategic places, mimicking the way marshes are formed naturally. The end will be "a gentle, vegetated slope" that will gradually increase in height, outpacing the rise of the sea level. It's worth noting that Dutch water-management experts were involved in elaborating the plan and in the extensive consultation with the public that led to its approval; the participative model developed in the Netherlands, which has also been used in the Bangladeshi plan for its delta lands, clearly was in play here. And, I'm pleased to note, the section of Highway 99 that skirts Mud Bay will get special attention.

All this is for a relatively limited area and the floodplains of two smallish rivers. What must be done on the floodplain of the Fraser River is a much bigger undertaking.

The Fraser is one of North America's great rivers. Rising in the Rocky Mountains of northern British Columbia, it flows 1,375 kilometres[28] to the Salish Sea, where its several arms fan out over a delta that was — and still is, where it has not been built over — farmland rich from centuries of sediments washed down from the mountains. The City of Vancouver sits to the north of the delta, its port facilities on the much deeper reaches of Burrard Inlet, a glacially formed fjord where Coast Range mountains plunge into the sea.

Flying into Vancouver from the east on a clear day, you get a panoramic view of this spectacular landscape. Sometimes the pilot may swing south near Mount Baker and other snow-capped peaks in Washington State before adjusting the course toward Vancouver International Airport. It sits on Sea Island, one of the big islands in the Fraser delta. As the plane swings around, you might see the wall of mountains to the north, some of them snow-capped, too. The airport, however, lies at sea level and is protected by a system of dikes like other stretches of shoreline in the Fraser floodplain. In total, 127 kilometres of dikes — many of them not built with rising sea levels and climate change in mind — guard the land from Fraser River flood waters or storm tides from the Georgia Strait.[29]

Among them are the dikes that protect the houses and farms in the municipality of Richmond and the extensive agricultural lands in the aptly named municipality of Delta. At the moment, Richmond has forty-nine kilometres of dikes and thirty-nine drainage pump stations that operate frequently during ocean storm surges and in the spring freshet season; that is, when melting snow in the mountains to the east swells the river. Plans are underway to raise the city's perimeter dikes to 4.7 metres over the next twenty-five to seventy-five years to stay ahead of climate change–induced sea-level rise."[30]

But dikes can only do so much, particularly since the land protected by them is actually sinking for two related reasons. First, as we've seen elsewhere, stopping floods also means stopping the addition of new sediment, and second, the sediment already deposited compacts over time. In fact, a recent report by soil engineers suggests that Lulu Island, the heart of Richmond, which is below mean sea level, would be 1.5 to 5 centimetres higher than it is today had it not been "protected" by dikes beginning in the early 1800s.[31]

As sea level rises, incursion of salt water into the surrounding land is inevitable, threatening the crops that grow there. So serious is the problem that Richmond is actively considering raising the level of the land on Lulu Island by bringing in soil from elsewhere, despite new problems that might be engendered in terms of soil fertility and contamination. Dredging up sediment from around the island might be an answer, since it, too, is largely sediment deposited by the river.[32]

The City of Vancouver is not faced with the task of protecting farmland, but it is very aware of the vulnerability of its coastline. Climate Central has done a simulation that shows what the centre of the city would look like with a 1.5°C increase in temperature: the many skyscrapers in the centre of the city would be standing, in effect, knee deep in water. Another simulation suggests that the house on Creelman Avenue in the Kitsilano neighbourhood that my sister once lived in would be under water, too. Already, the beach, which is a two-minute walk away, is sometimes submerged by surf during especially high tides. In January 2022, the parts of the eleven-kilometre-long seawall that curls around downtown Vancouver and ends at Kitsilano Beach Park were badly damaged by high tides linked to another serious weather event.[33]

In order to inform and prepare citizens about what will happen, Vancouver's city government has prepared some excellent materials: its booklet *Vancouver's Changing Shoreline* is exemplary for its explanation to citizens of what is at stake. The booklet includes a map that shows what parts of the city would be flooded by a major storm today were it not for the flood management measures already taken.[34]

≈

The beach at Kitsilano in Vancouver, B.C. It could be underwater in a few years' time.

Seattle is, to some extent, faced with similar problems, but since a good part of the city is built on hills, the threat of encroaching seas is different, except where low-lying land was chosen for development for specific reasons. Simulations by Seattle Public Utilities show the worst flooding forecast for the industrial and shipping area at the mouth the Duwamish River where it empties into Elliott Bay.[35]

Puget Sound, the most southern part of the Salish Sea, does not have a river as big as the Fraser flowing into it. Rather, a number of smaller rivers are its tributaries, but this is not to say that the rivers haven't been important in determining today's landscape — or that they won't be in determining tomorrow's. Some go as far as saying that the Duwamish, which rises to the south on the western slopes of the Cascades, is the reason why Seattle is there.[36] Certainly, the river can be used to study how rivers and their deltas are going to respond to rising sea levels.

For a number of years, my parents lived in West Seattle, a peninsula separated from the main part of the city by the Duwamish River. Crossing the bridge over the shipping channels constructed when the river was industrialized became a major part of the countdown to arriving at their house. My sister and her daughter would even sing, "Over the river and through the woods, to grandmother's house we go." No matter that there wasn't really a woods at that time, that they were never in a sleigh, and that there was rarely snow.

My father was working then for a car-import company located in the industrial lands that lined the much-abused Duwamish. As I remember, the landscape was pretty grey: concrete and asphalt and low buildings where myriad businesses did their thing. He took me in once to introduce me to his gang, treating me to lunch in a diner that catered to the people who worked in the industrial area. But I never found the neighbourhood a particularly attractive place. For scenery, we'd head to one of the parks that circled the northern part of the peninsula, like Alki Point. From there, we had an un-obstructed view across the water to Seattle and access to a beach that invited walking, wading, and building sandcastles. A seawall protected the roadway and the apartment houses that had been built along the road at the foot of the high ground. It was a good place to spend a sunny afternoon, and that's just what we did several times.

But even then I could see, without really trying, that a lot had changed in the neighbourhood since the first European-descended settlers had landed at Alki Point in 1851. Almost surely there had been woods on the hills be-hind the beach, and around the point to the east the Duwamish River had ended in marshes instead of today's cranes, docks, and asphalt. Members of the Duwamish Tribe had long fished, hunted, gathered, and celebrated near the river's mouth. Their longhouses, canoes, and other gear were well adapt-ed to a way of life that supported a healthy population of a thousand or more before the advent of European diseases like smallpox and flu. The ravages of disease weren't immediately apparent to the newcomers, though, because the diseases had swept through before the main wave of adventurers and mi-grants from eastern North America arrived in the mid-nineteenth century. What newcomers saw was land that seemed to be underpopulated and not well settled, where people were living in fashions that seemed strange and "undeveloped," to use a word from a later era. In short, a region that could be appropriated for the newcomers' purposes.

Within a couple of decades, they transformed it. One of the first activ-ities was chopping down the forests in order to ship lumber to San Francisco, which was growing rapidly as people poured in to join the Gold Rush in the in California foothills. That wasn't enough though; rather quickly, some of the newcomers began to plan how to tame the Duwamish River, to make it more amenable to shipping and manufacturing, and to farm the bottom-land through which it meandered. Before this transformation began, the Duwamish delta was a "beautiful plain of unrivaled fertility," according to one of the early settlers.[37]

By the turn of the century, however, great plans were underway to straighten the river, and to dredge its "foul" tidelands, which were "value-less for all purposes of navigation ... and generally unhealthy, unsightly and worthless for any and all purposes," according to Eugene Semple, thirteenth governor of the Washington Territory. Plans were also afoot to transform drastically the region's waterscape by building a canal connecting Lake Washington to the sound somewhat farther north. The aim of this and other projects was to "harness" Nature and make it more "productive," but they were undertaken with little understanding of how Nature worked.

One of the casualties was the part of the Duwamish floodplain that wasn't immediately reclaimed for industrial use. The low-lying land was "ideal for settlement," so the settlers cleared nearby forest to create open fields for their farms, ignorant of or uncaring about the forest's role in buffering high river flows and absorbing flood waters. As a result, "the pat-terns of erosion and deposition that created the fertile plains were disrupted. Without the forest cover, the plains were scoured, and their rich soil washed away."[38] Floods followed. Then came more armouring of the river's banks and a dam to channel its waters.

As the twentieth century progressed, the valley was filled in with indus-trial development — the Boeing plant had pride of place — and the canal was all too often used to dispose of waste. It wasn't until the century neared its end that the damage was realized and first attempts to right it began. By that time, "instead of engorged waterways the valley [was] filled with glutted highways and streets. Traffic jams [had] taken the place of logjams," as Alan J. Stein noted in his essay about the region's rivers.[39]

And the Duwamish River got dirtier and dirtier. It took concerted effort by ordinary citizens, the Duwamish Tribe, local authorities, and others to get the river classified as a Superfund site, the pollution category in the United States that qualifies a site for considerable federal aid. Beginning in 2001, a long-term program, which included removing contaminated sediments that harbour dangerous metals and chemicals, was adopted to clean up the river.

Twenty years later, there has been considerable improvement, in large part due to the removal of foul mud at several pollution hotspots. But work continues, and fish and other seafood that live in the river are still dangerous to eat, even though attempts to rehabilitate the river's banks have resulted in many places that look like a fisher's dream. The only exception is salmon, and that is because they spend only a short time in the river, when they are hatchlings and when they come back to spawn.[40] There's an irony — or a lesson, depending on how you look at it. Getting salmon back to the Duwamish required hands-on work by many. This has included actually carrying some fish bent on spawning upriver around stretches that they couldn't navigate because of the grade and the absence of a fish ladder — the very activity engaged in by Native Americans earlier.

A meander that was left after the ship channels were built one hundred years ago has now been restored. There, ruins of a Duwamish longhouse 1,400 years old can be seen, as well as glimpses of what the river was like before industrialization and farming arrived in a big way.

Rising sea levels will affect both these newly restored areas and the industrial and commercial enterprises that remain along the river.[41] Just as importantly, several residential communities that lie south of the greatest concentration of industrial and commercial uses are already being flooded regularly. Unfortunately, many of the people who are affected do not have the resources to protect their homes with elaborate defences. They are folks who, over the years, have moved there because of the neighbourhoods' proximity to both work and the river, and they include people from several recent immigrant groups. To meet their needs, the guide to safe fishing on the river has been printed in Khmer, Vietnamese, and Spanish, as well as in English.

The city has been working on improving drainage and building a pump station at a treatment centre to avoid discharging polluted water into the

Duwamish, but things will get worse. Projections show that large portions of the neighbourhoods will be underwater on a regular basis by 2100 when the sea level on Puget Sound is estimated to be as much as 1.2 metres higher than now.[42]

As for the beach at Alki Point where we had such good times, it is being flooded regularly by king tides — that is, the highest tides of the year boosted by weather conditions. In 2021 work was begun to build a 150-metre stretch of seawall to protect the roads and other infrastructure threatened by the waves; when sewer lines were first laid along the shore, no one thought they'd be endangered by rising seas.[43]

But raising the level of the land, as is suggested for Lulu Island, or building new seawalls and replacing ones that are being undermined — are these the answers to rising sea levels? Not really. Nor will half-measures bring under control the increasing threat from wild weather, like the days and days of torrential rains in the Salish Sea and its uplands in the fall of 2021. What will be needed are both low-tech approaches and a couple of very high-tech ones, as we'll see in the next section.

PART 4

WHERE WE GO FROM HERE

MUSICAL INTERLUDE

The French say that *la vie n'est pas un long fleuve tranquil* — life is not a long tranquil river. There's a lot of truth in that, particularly as the stakes of climate change grow higher. But as with much else, music — one of the glories of civilization — sometimes expresses what we're up against better than any other medium. That's why the next musical interlude must be the fourth movement of Beethoven's Symphony No. 6. Beethoven, who loved the countryside, called the symphony as a whole *The Pastoral*, and this movement simply "The Storm." A rousing performance of "The Storm" by the Warsaw Philharmonic Orchestra can be found on their YouTube channel, Filharmonia Narodowa.[1]

In the fourth movement, Beethoven interrupts the jolly, playful atmosphere of most of *The Pastoral* with the sudden arrival of bad weather. It begins with the double basses and cellos churning against each other and the percussion rumbling away mightily. Piccolo, timpani, and trombone cut across the music the way lightning cuts across a sky darkened by a thunderstorm. Then, almost suddenly, the storm subsides and the fifth movement begins immediately with a lighter, happier melody which Beethoven called "Shepherd's Song — Happy, Thankful Feelings After the Storm." That ending is something to keep in mind as we struggle to find our way through the gathering storms of climate change. Maybe we can make our way out, too.

8

WHAT CAN BE DONE – WHAT WILL BE DONE?

Covid-19 took up so much space in our lives for so long that by the late summer of 2021, I found myself surprised when people began to turn their minds to what they'd put aside during the health emergency. The Indonesian government was still grappling with very high case numbers at the end of August 2021 but nevertheless decided to move ahead with the plan to transfer the capital from Jakarta to East Kalimantan. The announcement came on August 27 when President Widodo visited the new site to inspect a toll road that is supposed to be a major access road for the new city, to be called Nusantara. "The new capital agenda will proceed as planned," he proclaimed. Two days later, the head of the National Development Planning Agency added that construction will take a couple of decades, not the few years originally planned.[2]

That clinched it. I decided definitively that there was no point in trying to get to Jakarta anytime soon, even if Covid-19 restrictions might allow it in a few months. Best to forge ahead with this project, even though many questions remained unanswered in Indonesia: new appropriations for preliminary work hadn't been approved yet, and what to do about rising seas in the old capital wasn't being talked about.

So, life goes on — or it does for those of us who were not among the millions tragically killed by the pandemic. The clear skies of the lockdowns the world experienced in March and April 2020 had long passed. By May 2022, the amount of CO_2 in the atmosphere had reached levels 50 percent higher than those before the Industrial Revolution,[3] and "exceptional" weather was becoming the norm. Ten years from now, we may look back at Covid-19 as something of a dangerous diversion that kept us from thinking seriously about what we collectively would do about climate change and rising sea levels. We may also lament losing the opportunity to shift from fossil fuels that the war in Ukraine almost, but not quite, forced on the world. Instead of grousing about rising gasoline prices, we could have insisted that alternate paths be followed that would lead to more reliance on renewable energies of all sorts. First and perhaps foremost, countries could have begun taxing the windfall profits that fossil fuel companies earned and started to plough the money back into green initiatives.[4]

But that would require collective action, and the term *collective* is a loaded one. The reaction of a significant number of people in North America to public health measures designed to control the pandemic — the protests of the anti-vaxxers and anti-maskers — shows that getting folks to rally to the cause is not going to be easy.

The first question that must be asked is this: Can we turn back the tide, figuratively and literally? All the data, all the studies suggest that given the current pace of greenhouse gas emissions, we will surpass an average temperature increase of 1.5°C degrees, unless we — and by this, I mean people and countries all over the world — do a number of things.

In the spring of 2022, the Intergovernmental Panel on Climate Change (IPCC) declared in a much-anticipated report that it isn't too late if we all pull together.[5] When the report was released, IPCC chair Hoesung Lee said, "We are at a crossroads. The decisions we make now can secure a liveable future. We have the tools and know-how required to limit warming." He added, in words designed to hold out enough hope to keep the world from giving up completely, "I am encouraged by climate action being taken in many countries. There are policies, regulations and market instruments that are proving effective. If these are scaled up and applied more widely and equitably, they can support deep emissions reductions and stimulate innovation."[6]

But that optimism has to be tempered by the difficulty of getting the world truly committed to cutting emissions enough. What countries have said they'll do so far is a topic we'll return to later. Needless to say, some ideas are simpler to accomplish than others.

Trees and Other Carbon Sinks

As noted before, there's a body of research that suggests that the Little Ice Age, that great dip in temperature that occurred between the fourteenth and late eighteenth century, may have been caused in large part by the regrowth of forests following the crash of populations caused by the Black Death, European invasions by Genghis Khan and his followers, and the introduction of diseases like smallpox, measles, and malaria by Europeans into the Western Hemisphere.

Based on records from the first years of Spanish conquest, it looks as if the Indigenous populations in Central Mexico plummeted from something like 25.2 million in 1518 to 700,000 in 1623.[7] The chronicle of the first European voyage down the Amazon reported many settlements along the river's upper reaches, and several very large towns farther down; evaluation of the chronicles suggests a population of around five million. The contrast with the empty shores that later explorers found cast doubt on the accuracy of the first reports, but recent archeological discoveries indicate they were right on the money.

In Canada, population estimates at time of European contact range from two hundred thousand to two million, but when Samuel de Champlain sailed up the St. Lawrence, he found no sign of the prosperous villages along the great river's valley that Jacques Cartier had reported sixty years before. Part of that population displacement may have been due to intertribal warfare, but the dramatic depopulation points to epidemics raging through Indigenous communities. The fields were left to grow fallow; far fewer fires were set to manage woodlands; settlements were deserted. As a consequence, trees grew back, sopping up carbon dioxide. The net result of this decline in a major greenhouse gas appears to have been a significant worldwide decline in temperatures.[8]

The IPCC 2022 report alludes to the role of forests as CO_2 sinks in its recommendation that people reduce their consumption of meat, particularly beef. Not only do the burps and farts of cattle produce methane, a powerful greenhouse gas, but also great tracts of forests in places like the Amazon basin are being cut down to provide pastures for the beasts.

The take-home lesson from this is that increasing forest cover and decreasing deforestation could make a major difference in our future. The 2021 floods around the Salish Sea were made worse by deforestation and forest fires upstream, because water that would have been slowed by trees and absorbed by tree roots had nothing to stop it flowing downhill. The same dynamic has been playing out around Jakarta, as more and more of the forests in the Ciliwung's watershed have been cut down in the last fifty years.

The obvious corollary to this is that we should replant forests. To be really effective, however, reforestation projects should not be the simple one-species plots that have been the norm to date.[9] All forests are not equal in the amount of CO_2 sequestered, since there is a difference in the carbon uptake among tree species.[10] Furthermore, when it comes to biodiversity, you can't equate a plantation of oil palms to a natural tropical forest. Burning the latter down to make the former does damage of many sorts. For example, on some days in 2015, smoke from fires set in Indonesia so that forests could be replanted in oil palms produced as much greenhouse gases as the entire United States was emitting at the time.[11] Furthermore, when a tropical forest is burned, the peaty soil on which it stands is exposed to the air and begins to degrade, releasing carbon dioxide and methane that has been stored there. This amounts to a double blow to the Earth's carbon budget.[12] Even logging for lumber is not as bad, since a major portion of the trees cut down to build and manufacture things remains as a carbon sink. In fact, it may be that cutting down forests in order to build with wood might help right the carbon dioxide balance, as long as the forests are replanted wisely. That is because both concrete and steel, which have been the go-to materials for construction over the last century, have an enormous carbon footprint. Making each of them emits more CO_2 by weight than the material itself weighs: the ratio is 1,028 kilograms of CO_2 per 1,000 kilograms of cement produced, and 1,850 kilograms of CO_2 per 1,000 kilograms of steel.[13]

Dry wood, on the other hand, is about 50 percent carbon, and the trees grown to replace those cut down continue the cycle by storing more carbon. Mass wood — thick, compressed layers of wood that can be used in constructing tall buildings — is being touted as a way of reducing global CO_2 emissions by 9 percent if the material were used for half of new urban construction.[14]

But forests in the countryside are only part of the equation. Trees in urban areas matter a lot too.

So, it's a hot, muggy Montreal summer afternoon and I get off the bus two blocks from my street. The sun glares off the cars stopped at the traffic light at the corner of avenue du Parc and St-Viateur. The cab drivers at the taxi stand are huddled in the shade of the one anemic tree at the intersection. I'm feeling beat: I'm not made for hot weather, and I hate to think what it must be like to have a heat wave like they're experiencing around the Salish Sea and inland a ways. But I trudge onward, until I come to my street and turn up it. There are street trees here, some planted more than one hundred years ago when the neighborhood was platted, some much newer and smaller but evidence of a commitment to provide islands of shade and coolness in the city. This too is a way to cut down on effects of climate change because not only are these trees soaking up CO_2, they're also lowering the ambient temperature and making it less tempting for folks to run their air conditioners.[15]

Planting trees in cities definitely should be on our to-do list, but less natural ways of pulling carbon out of the air usually generate more excitement in the media. Take the Orca carbon capture plant in Iceland, which is supposed to be able to suck four thousand tonnes of CO_2 out of the air annually and inject it deep into the ground, where it will become rock. (The name, by the way, has nothing to do with the orcas of the Salish Sea, which seem to be doing so poorly faced with climate change. This Orca comes from the Icelandic word for energy, *orka*.) When the installation started operation in November 2021, media around the world featured it. The carbon that it will sequester is the equivalent to that emitted by about 870 cars — or the production of more than four thousand tonnes of cement — in other words, a drop in the bucket compared to what would be needed for carbon sequestration to make a difference. What is more, the process requires a lot of electricity. That makes it much less attractive as a solution to our CO_2 problems in places unlike Iceland, where the energy comes from harnessing the heat generated by the island nation's volcanic underpinnings.[16]

I suppose I shouldn't be surprised at the splash the Orca project made, because it's often more impressive to talk about technical solutions. Chief among them are projects to protect places from what's happening now, and what is bound to happen in the future. Note that they do little or nothing to cut down on greenhouse gas emissions, and thus are not tools to turn climate change around.

The Thames Barrier

Take, for example, the Thames tidal barrier, which is one of the most effective high-tech constructions designed to control rising seas. It was prompted

The explosion of gunpowder magazines near the Thames in 1864 nearly destroyed the embankment.

by London's experience during the extreme weather of 1953. The same storms that hammered the Netherlands, killing more than 1,800, also hit southeast England, with a storm surge roaring up the Thames to overtop the embankments that line the river's course. Something had to be done to prevent similar disasters in the future, authorities agreed. Just raising embankments and dikes wouldn't be enough; far more cost effective would be a barrier to stop storm surges. At the time, the prospect of rising sea levels was not on the horizon. It was enough to realize that tides and storms in the Thames Estuary have long been serious threats to the people who live and work on the Thames floodplain, all several million of them.

Despite the concern, it took nearly thirty years for the barrier to be designed and built. It "spans 520 metres" of the River Thames and "protects 125 square kilometres of central London."[17] Operational in 1982, the barrier has ten steel gates, each weighing 3,300 tonnes. Six of them rest on the river bottom under normal circumstances, allowing for tides to ebb and flow and for ships and boats to travel up and down the river. When storm surges are forecast, these gates can rotate up from the bottom of the river, forming a wall as high as a five-storey building. Four other gates on either side drop into the river to complete the barrier. When in place — a process that takes about an hour and a half and is usually started at low tide — the barrier holds back the surge until the tide ebbs and the surge passes. Then water is allowed to flow under the side gates until the level on either side is equalized so that the rotating gates can return to the riverbed.

As of June 2021, the gates had been used 205 times: 114 times to resist tidal surges and 91 to protect against a combined threat from tides and floods. In the future, the Thames Estuary 2100 plan proposes to maintain existing structures — the Thames Barrier was originally designed to last until 2030 but is now thought to be usable with modifications until 2070 — but also to raise some embankments and smaller barriers as time passes and climate change advances. Among these — and we'll return to this later — are ways to make the river more accessible by introducing more soft solutions to the problems of flooding and tides.

Living with storm surges, rising tides, and floods is nothing new on the Thames. Over the last few decades, archeological research in the estuary has turned up evidence that for at least three thousand years a relatively large number of people lived there, subsisting — even thriving — on the tidelands' resources. Fish, waterfowl, reeds, and other plants were there for the taking; salmon were plentiful until modern times, and, as we've seen around the Salish Sea, salmon can be very important to a society.[18] The picture that is emerging is one of societies sophisticated enough to build wooden trackways across marshes and which left remnants of gravel passages in other places. Most of the finds are buried under layers of peat and other organic material that preserved them over millennia — and which, significantly, suggest that the land was slowly buried under silt deposited by floods and rising tidal flow.[19] If this sounds like what must have happened in Doggerland, that is no accident, because what is now the Thames Estuary is in effect an extension of that ancient, now-submerged country.

In fact, the path the main channel of the Thames takes to the sea seems not to have changed much over the millennia since the final flooding of Doggerland. Rather, in classic river delta fashion, where it neared the sea, the river divided into many meandering streams that were strongly affected by tides. So sophisticated are some of the early embankments along these channels that historians and engineers long thought that the first of them must have been constructed by those great engineers, the Romans, during the nearly four centuries when they held Britain. Archeologist Flaxman Spurrell realized in the 1880s, however, that the Romans probably wouldn't have needed to put them up to constrain the river, because they built their headquarters on sites probably chosen because they were on higher ground.[20]

The first embankments clearly predate the Romans, however, and did not require advanced engineering techniques. They began as small-scale efforts to protect fields, and as sea levels rose and the tides reached farther up the river, their owners built the embankments higher. Spurrell notes that a timeline can be deduced from such things as "marshes of to-day [in which] houses stand, and broken glass and bones and other rubbish would indicate the date they were abandoned." In other places, "Roman pottery is so plainly detected that we know by it what was the level of the Roman period." Recent dating by carbon-14 and other methods corroborates Spurrell's conclusions. (It should be noted, however, that he suggested that the main reason for the incursion of tides higher up the Thames was the land sub-siding. Today, it seems clear that, while there may have been some rebound effect going on after the end of the last ice age, the major force behind these intrusions has been and continues to be the rise in sea levels themselves.)

Along the way, people began to reclaim marshland, using techniques employed all over the world: enclosing the marshlands and installing sluice gates to allow fresh water to escape and to keep out the tides. Spurrell writes that in the district the process of "inning" (bringing the marsh "in" from the danger of tides) was begun "from the hard land, and banks are carried out a certain distance, returning to the dry land at some other place; then from some point of that line other essays are made until a large area is enclosed." This sounds very much like the way Dutch polders and Canadian *aboiteaux* were built.

Spurrell continues: "Many writers are impressed with the 'mighty,' 'stupendous,' or 'vast' embankments which keep out the water of the river [but there] is no need for such expressions.... The height to which we see them now rise, is the gradual increase from slighter banks which costs but little exertion, although regular attention." He adds that even in the most difficult situation, two hundred acres were inned in two years, "with an average employment of 30 workmen."

Well into the eighteenth century, drawings and paintings of the river's course show many low banks in the downstream reaches. At the same time, seawalls and the raising of banks in central London squeezed the river, which meant that the water was forced to run faster through the narrow sections. This in turn scoured the bottom, making it deeper and therefore easier for larger ships to navigate. When there was a breach in an embankment, frantic work had to be undertaken to close it before the tide mounted. One of the most dramatic failures occurred in 1864, when two gunpowder magazines on the south side of the Thames between Erith and Woolwich (now the district where the Royal Arsenal is located) exploded.[21] An account written shortly afterward gives a taste of the danger of the explosion and of the techniques of communication and construction available at the time:

Fortunately, the tide was out at the time of the explosion; and it was necessary to close the gap before high water at one o'clock to prevent an inundation that would have caused incalculable damage. Mr. Lewis Moore, an engineer connected with the Metropolitan Board of Works, who resides at Erith, sent a message to Mr. Webster, the contractor at the main drainage works at Crossness; and he, with praiseworthy promptitude, at once dispatched a strong force of 350 navvies, who set to work with a will. Information of the danger was telegraphed to the authorities at the Horse Guards, who, in reply, telegraphed an order to Woolwich that every available man in the garrison should be sent to repair the breach. General Warde, the Commandant, at once dispatched 1500 men, marines, horse artillery, and engineers, with 2000 sandbags.... Part of the troops lashed beams and spars together and floated them in front of the breach to act as a breakwater and prevent the swell of passing steamers from washing away the newly-formed embankment. Others were stationed on the place where the embankment was to be raised: from these double files were extended to another large party in the rear. Hundreds of sandbags were filled with earth by the party in the rear; then, as fast as they were filled, they were passed to those at the front, by whom they were piled. The moment of danger was about one o'clock. But by the most energetic and praiseworthy exertions on the part of all engaged ... the temporary embankment was raised rapidly enough to keep pace with the rising tide. At high water the river was found to be oozing through in several places, and great apprehension was felt for the result. There was no despairing, however: the navvies redoubled their exertions, and fresh detachments of troops took the place of comrades wearied by strenuous labour; so that the dangerous period was got over without any serious influx, and before the tide had

Finished in 1982, the Thames Barrier has guarded London from storm surges successfully.

fully ebbed the breach was so strengthened that there was no immediate danger of an inundation.

Whew! That was a close call. There have been others, but in general the worst has been averted.

This embankment building reached its peak in Victorian times, when three separate grand embankments were constructed in central London. These found their echo, as we've seen, in Shanghai's the Bund and Buckland's Bund in Dhaka. Today, say some authorities, the Thames from the Teddington Locks to the North Sea is, in effect, a canal, and were its banks removed, the river and tides would flood a vast area.

But today, to walk along the Thames in some places on a nice afternoon is to miss entirely the drama of the Thames's continual battle of tide and flood. The towpath that runs between Kew Bridge and Richmond (only one stretch of the much longer trail that begins in the Cotswolds and goes nearly to the sea) is bucolic. On one side lie the Royal Botanic Gardens at Kew, the largest botanic garden in the world. On the other, at the bottom of a sloping embankment, the river rolls on. It rises and falls with the tides even here, but on a day like the one when I walked it a couple of decades ago, the river looks gentle. Across from Kew sits Syon House, owned by the Duke of Northumberland but now open to the public. The current building dates back to the sixteenth century, but long before, it was the site of a country monastery. A large meadow leads down to the river. Regularly flooded by high tides, it is one of the few places on the lower Thames where walls and banks don't constrain the river. Overhead, a steady procession of airplanes take off and land at nearby Heathrow Airport, noisy reminders of the changes we are wreaking with our addiction to speed and fossil fuels.

But for the moment, it appears that the Thames Barrier and other measures are keeping the waters at bay, to make a bad pun. This doesn't mean that the problems of flooding in major storms are solved, though, since torrential rains can back up before they reach the river, as happened twice in 2021.

MOSE, Venice's Barrier

Nevertheless, the British have reasons to point with pride at their accomplishments, now and in the past. As for Venice, the case is much less clear. Shortly after the Thames Barrier was finished, planning began there for a three-pronged barrier project to protect the city of canals from tidewaters. Now, nearly thirty years later, MOSE (Modulo Sperimentale Elettromeccanico, or Experimental Electromechanical Module) — the name was chosen with a nod to the Biblical story of Moses who, after all, supposedly parted the waters — has begun lumbering into service after decades of scandal and years of engineering challenges. As of October 2021, it had twice saved the city from serious flooding due to the *acqua alta*, the extreme high tides that plague the city, but it was clear that the half-a-billion-dollar project was not going to stop "ordinary" high tides, which hamper life in the storied city.

The idea behind MOSE is different from that which led to the Thames Barrier. Strictly speaking, Venice isn't a river city like London, although its origins are tied up in salt marshes and rising sea levels, too. Its backstory has been teased apart by scientists studying cores bored deep into the sediments of the Venetian lagoon and on the higher ground that surrounds it. The region is quite active, seismically speaking — think of all those earthquakes that periodically lay waste to sections of Italy. It also has been affected by the duel between sediment and sea as rivers flowing down from the Alps and the Apennines have left their loads of sand, gravel, and mud over millennia. In the ordinary run of things, these sediments would been enough to fill a low-lying swale, but sometime around six or seven thousand years ago the rising waters invaded it, producing Venice's famous lagoon.[22] The embayment wasn't unique; there were several along the coast, but none became the site of a city whose shadow extended as far away as the riches of the Ocean Isles on the edge of the Pacific Ocean.[23]

The city itself dates from the chaotic period when the glory that was Rome faltered and fell. In the fifth century CE, refugees from "barbarian" forces besieging Rome took shelter on several of the 118 islands in the lagoon. The settlement became a city state, electing its own rulers, and later a republic whose influence was felt around the Adriatic. More importantly, because of its strategic location, it came to control much of the trade around the Mediterranean basin and beyond. Textiles, spices, and glass were among the materials exchanged by Venetian merchants; Marco Polo was only one of many who voyaged to the Far East. The desire to circumvent the Venetian

Venice and its canals. *The Entrance to the Canal Grande at the Punta della Dogana and the Santa Maria della Salute* by Giovanni Antonio Canal.

commercial dominance lay behind the efforts of first the Portuguese and the Spanish and then the English, Dutch, and others to find an alternate path to the riches of East Asia. A measure of the long journey of nutmeg, that signal spice of the Ocean Isles, can be seen in its use in Italian dishes, like Marcella Hazan's classic recipe for Bolognese sauce, which calls for gratings from a whole nutmeg.[24]

Originally, the islands in the lagoon were relatively small, and a way had to be found to house the people and commerce. Among the methods used was extending buildings into the lagoon's waters by driving wooden stakes into the muddy bottom and then constructing platforms on top. Many of the original pilings remain, preserved in the mud and unoxygenated water. The spaces between became canals, and the augmented islands were crowned with buildings, many of which are considered architectural marvels.

The lagoon, however, has only three access points, through which tides flow twice a day and water from the rivers exit. Much of the original tidal marshes of the lagoon's curving northern shores have been built on for industrial uses. The combined result means that the lagoon's waters have long been polluted, while the remaining marshes for centuries were breeding grounds for malaria-bearing mosquitoes. Nevertheless, the city has become one of the most frequented tourist destinations in the world — some twenty million visited in the last pre-pandemic year, 2019.[25] And this despite the fact that guidebooks suggest bringing wading boots because places like the square in front of Saint Mark's are regularly flooded. Not unlike Jakarta, Venice has sunk quite a bit due to land subsidence, so the problem of rising sea levels is compounded by lower land levels.

I mentioned Amitav Ghosh, the Bengal writer, in chapter 6, "Dhaka and the Sundarbans." His novel *The Hungry Tide* contains extremely evocative writing about the vagaries of great river deltas. The themes he dealt with there, including the problems of flooding, saltwater intrusion, and pollution, are ones that have preoccupied him since. He wrote a non-fiction book, *The Great Derangement*, that bemoaned the fact that few writers were attempting to set their stories in a world where climate change menaces.[26]

In order to help fill the void, more than a decade after *The Hungry Tide* was published, he produced a sort of a sequel, *Gun Island*. The book connects the Sundarbans and Venice in a way that at first seems far-fetched and then becomes an example of the way the world is connected and has been connected for centuries. After adventures in the Sundarbans, the narrator finds himself involved in a search for the truth behind a legend of a Bengali merchant in Venice (Ghosh is, of course, aware of the literary resonances). He tells of visiting an Italian friend who lives in an apartment off the Grand Canal, where rising sea levels are more than a distant threat. In the early 2000s, when he last visited, one entered the building from its private dock on the first level, but on his next visit, twelve years later, the residents usually use an entrance in the back, through a somewhat higher garden, because even with gangways the lobby is flooded much of the time.[27]

But that is not the only surprise of this visit: he is astonished to discover that many of the young men working in the tourist trade and on construction sites in Venice are Bangladeshi. On a day of *acqua alta*, he encounters many of them tending gangways, selling boots and galoshes. One of them says, "We earn well on days like this. For us it's like home — we're used to floods." Presumably, on days of ordinary flooding, business is less brisk.

Ghosh says that climate migrants like these young men have become essential to the functioning of much of the "developed" world, even though these displaced people are not welcomed when they attempt to migrate. Sometimes, indeed, it takes a miracle for them to get in — but for more about that, you must read the novel.

There is no miracle behind MOSE, though, despite the passing nod its name gives to Moses's exploit. Its seventy-eight mobile barriers are designed to come up from the bottom of the lagoon to block the incoming waters at three channels between the lagoon and the Adriatic. Plagued by corruption and cost overruns, its construction took years. It was supposed to be completed a decade ago, but only became fully operational at the end of 2021, although the gates were raised for the first time in the summer of 2020. The following fall, they were credited with saving Venice from major flood damage in November, but in December 2021 the gates did not activate because of a faulty weather forecast. Tidewaters rose to a high of nearly a metre and a half above sea level on the afternoon of December 8, flooding St. Mark's Square, the lowest point in the city.[28]

Human error on the part of forecasters, of course — even MOSE's critics agree. But one could say that this whole climate mess we're in is just another gigantic example of human error all over the world.

It should be noted that MOSE has been the focus of all the attention surrounding attempts to solve Venice's flooding problems, but efforts are also being made to raise embankments and buildings in the heart of the city, while on the northern perimeter of the lagoon, not only is pollution being cleaned up, but also some of the formerly dangerous salt marshes are being re-naturalized in an attempt to provide soft barriers for the rising waters.

Do We Need to Go Dutch?

This soft mitigation has also become a very important element in the struggle against rising water in the country that can be said to have invented high-tech defences: the Netherlands. The rethinking of the best way to go forward began in 1995, about the time that the last of the big Dutch dam projects designed to keep seawater out of the country were being completed. There had been talk for some time about integrated water management; that is, looking at the problems in a way that included many approaches. However, the threat of river flooding was not taken really seriously until extreme river water levels in the 1990s nearly caused dike breaches and led, at one point, to the evacuation of a quarter of a million people and a million cattle. What was needed, the Dutch government decided, was more "room for the river."

The 2.2-billion-euro plan was begun in 2006. Thirty-nine locations were selected for projects to divert and absorb the water. Instead of building over canals, as was being done in some urban areas, and increasing the amount of hard surfaces like paving and parking lots, the idea was to allow the water to drain naturally. (If this sounds rather like China's "sponge cities," it's no accident, as Dutch consultants advised the Chinese government.[29]) Among the methods used were building flood bypasses, making floodplains deeper to hold more water, and relocating dikes. Many of the places where the rivers could spread out during high water were converted into recreational spaces that in ordinary times are used for pleasure. Collateral projects included teaching kids how to swim wearing shoes and clothes; the emphasis has to be on escape by water if the new defences fail because, as the mayor of Rotterdam explains, the nine hundred thousand people in his city could never be evacuated by land in times of severe river flooding.[30]

So far, the scheme has worked. There was some flooding in the Netherlands in July 2021, but post-mortems show that the measures put in place over the last two decades were largely successful in managing record-breaking rains that caused havoc for the Netherlands' neighbours. Even though a project to control flood waters on the Maas River (called the Meuse in France) hadn't been completed and the amount of water rushing to the North Sea reached near record levels, the region suffered much less damage than it did in the brutal floods of 1993 and 1995. True, there was much damage where a dike broke, but Dutch authorities say that "room for the river" tactics showed their worth.[31] It was quite another story in Germany, Belgium, and Luxembourg, where inadequate warnings and insufficient water control led to massive floods and the deaths of at least 199 people.[32]

What the Dutch have done when it comes to high tech is also extremely successful. Early in the twentieth century, the country began the Zuiderzeewerken, mentioned earlier, which involved damming a large, shallow inlet of the North Sea and converting the bottomland into agricultural land. This included constructing the Afsluitdijk, a thirty-two-kilometre causeway that is ninety metres wide and more than seven metres above sea level. After the disastrous flood of 1953, the focus shifted from reclaiming land to protecting what had already been reclaimed. Today, there are thirteen storm surge barriers, including the Oosterscheldekering, where the Scheldt River meets the sea in the southwest of the country. Initially, it was to be completely closed, but, after public protests, it now contains four sluice gates to allow movement of water between the estuary and the sea. Between 1986, when it was completed, and 2020, the full barrier was closed twenty-seven times when storm surges of more than three metres were forecast. As a plaque on the barrier says, *Hier gaan over het tij, de maan, de wind en wij* (Here the tide is ruled by the moon, the wind and us [the Dutch]).[33]

The last of these Deltawerken projects to be completed is the Maeslantkering, a barrier that protects the gateway to the port of Rotterdam, a city that is 90 percent below current sea level. At first, the plan had been merely to raise dikes protecting the shoreline, but given the projections of rising sea levels, to do this adequately would have meant the destruction and removal of many existing buildings and portions of old towns. Instead, what was installed is a system made up of two gates, each the size of a toppled-over Eiffel Tower, on either side of the Nieuwe Waterweg. The gates swing closed automatically when a storm surge of three metres or more is forecast. So far, it has been closed only twice during a storm, in 2007 and 2018.[34] However, the gates are tested every year, and, because of recent predictions of the extent of sea-level rising, they will be reinforced to stand two metres higher.

Storm surge barriers are on the drawing boards for several cities elsewhere in the world. The U.S. Army Corps of Engineers developed a plan to protect New York from hurricane damage that would involve building several artificial islands in New York Harbor that would be connected by 9.6 kilometres of retractable gates, rather like the Thames Barrier. While the plan was put on the back burner by the Trump administration in 2020, the city is moving ahead with other means to mitigate the damage produced by severe storms.[35] These will include building smaller seawalls and elevating a park along the East River by 2.4 to 3.0 metres so that they form a barrier to high water. Some of these innovations, hopefully, will also aid in diminishing the effects of heavy rainfall, which is predicted to be more common in coming years. Certainly, the storm surge barrier that was proposed after Hurricane Sandy would not have prevented the flooding of the subway and basement apartments that led to the deaths of forty-five people in September 2021, when torrential rains fell on the city.[36]

But all of these projects pale in comparison with one proposed quite seriously by Sjoerd Groeskamp, a young Dutch oceanographer, and Joakim Kjellsson, an equally young Swedish meteorologist. Their plan: wall off the North and Baltic Seas from rising oceans and catastrophic storm surges with dams connecting France, England, Scotland, and Norway.

Sounds bizarre, and when I came across the idea in early 2021 in a YouTube video made by the Azerbaijani vlogger Shirvan Neftchi for his channel, CaspianReport, it seemed so far out, so much like a science fiction story, that I put it out of my mind.[37] Then I stumbled on articles in the *Guardian* and *Science Daily* about the proposal and realized that a number of other scientists had decided it was worth thinking about. The prestigious, peer-reviewed journal *Bulletin of the American Meteorological Society* had published the pair's proposal in July 2020, in an article titled "NEED: The Northern European Enclosure Dam for if Climate Change Mitigation Fails."[38]

The idea is a sort of shock therapy for the world, the article's authors say. Groeskamp — wearing a T-shirt emblazoned with the name of his favourite reggae/ska band — explained to me in a Zoom interview that their initial idea had been to use *when* rather than *if* in the title, but their editor said that it was just too pessimistic.[39] He agrees that it is very important not to give up hope that efforts to reverse CO_2 emissions can save us from the worst of climate change. But the choice we have before us is very clear, he says: we have to do everything we can to change the path we're on, or the massive enclosure dam — a 637-kilometre-long structure — could be the most cost-effective way to protect the fifteen northern European countries that would be behind it.

Microturbulence in the oceans is Groeskamp's main research interest, but he began thinking about NEED in 2017 when he was preparing to return to the Netherlands after several years in Australia and the United States. "My wife and I wondered, should we buy a house?" he told me. The research institute where he would be working is on Texel Island in the Wadden Sea, one of an arc of islands that help protect the mainland of the Netherlands. "We started looking at the maps and it didn't sound like a good idea," he said. Scenarios for sea-level rise go as high as fifteen metres by 2300, and the short-term forecast is for a metre or so in the next couple of decades. "We'd be living in a bathtub if the dikes broke," he said. His conclusion: climate change must be reversed, or heroic efforts will be needed to protect not only Texel Island but most of northern Europe. More of the same — higher dikes and bigger storm surge barriers — could only do so much, particularly if the Dutch held out but Germany and Belgium decided to retreat from the coasts and leave them undefended.

So Groeskamp did some back-of-the-envelope calculations — the idea became "my hobby," he said — and decided that the dam might be possible, given the cost of building protective dikes and barriers compared to that of an enclosure dam. Since their paper was published, he and Kjellsson, who works for a research institute in Germany, have received many messages of interest from around the world, some from people who are enthusiastic about the idea already, and others from those who are at first shocked and then realize how much is at stake.

For example, NEED would mean that eventually the North Sea and the Baltic Sea would fill with fresh water coming from the many rivers that drain Europe. "My biologist friends are furious about that because whole ecosystems would be completely destroyed," Groeskamp said. "I tell them that if they are so angry, they should really work to stop climate change."

Opening up the conversation about the problem is all to the good, he insisted, not least because NEED would address only the problems of northern European countries, but rising sea levels are going to affect every country with a seacoast. Their paper also briefly sketches enclosure dam proposals for the Mediterranean, the Irish Sea, the Sea of Japan, and the Persian Gulf. But for geographic reasons, many other regions — the Sundarbans among them — would have no possibility of building this kind of defence even if the money were found to do so.

"We have ten to thirty years in which to act," Groeskamp told me, "but if we don't, at least this idea is out there, and people will have been thinking about it." The need (pun intended) for major action will remain even if we can hold temperature rise to 1.5 or 2°C and then get it to stabilize, he said, stressing that the problems of rising seas are with us for the long haul. Because of the temperature increase, the water from the melted ice caps and glaciers will have already gone into the oceans, and the very volume of water will be greater.

Can we do it — can we stop climate change? Well, Groeskamp is cautiously optimistic that if we throw every tool we have at the problem of controlling greenhouse gas emissions, a sort of liveable equilibrium can be achieved. He and his wife are confident enough in that to have bought a house, in fact, although he admits than in ten years or so they may have to re-evaluate the situation and move.

Bringing Out the Big Guns: Going for Nuclear Power

Recently I've come across several YouTube videos debunking the NEED idea, pointing out all its weak points, not the least of which is getting all the countries around the North and Baltic Seas to agree to a project like that. The United Kingdom has opted out of the European Union, after all, and Russia, whose jewel Saint Petersburg lies at the head of the Baltic, is to many a pariah state as a result of its invasion of Ukraine. All true, but what I didn't expect to see on YouTube was Bill Gates, one of the founders of Microsoft, shilling his latest book.

Gates, of course, knows how to use electronic media. On his YouTube channel, he posts his thoughts and research about a number of topics. He is pretty good at breaking down complex subjects into understandable bites. When I was working on my book about concrete, I found the information he brought forward about the great increase in Chinese concrete consumption a game-changer for my thinking. But I'd never before seen him advertising something in one of those annoying twenty-second spots that pop up before the start of the YouTube video you want to see. There he was, though, wearing a light brown sweater over a dress shirt with the collar unbuttoned, and looking like a kind but nerdy high school teacher.

What he was talking about was his book *How to Avoid a Climate Disaster*. I take it as a measure of just how concerned he is about where we're headed that he was making the pitch in this way.

In the book, he begins by reviewing why we're in a CO_2 crisis, before marshalling evidence about how the crisis might be rectified. Zero carbon is the goal, and the ways to get there are many, he says. Carbon taxes can be extremely effective but won't be enough. Research and development into new techniques for such approaches as storing energy from wind and solar power will be essential. So will sucking CO_2 out of the air for sequestration underground. We all must use less fuel in doing the things we must do. Cutting down on eating meat would help. Everything that can be electrified should be.

He had me up to that point. There wasn't anything new to anyone who'd been paying attention to the climate change debates lately. But then he began talking about the need to reconsider the use of nuclear fission — and later, nuclear fusion — to generate electricity.

Now, like most of the generation that Gates belongs to, I am familiar with the hazards of messing with the atom. I grew up under the threat of nuclear war, with duck-and-cover drills in classrooms, and grown-ups who were just as spooked as we kids were at the idea of the atomic bomb dropping. I remember being hysterical one sunny afternoon when I was about six, because there were bombers flying above our heads in the little town in eastern Washington State where we lived, practising landings and takeoffs or perhaps just showing off. The grown-ups assured me that nothing was going to happen, that it was highly unlikely that there were any bombs within several thousand miles (surely an exaggeration, I realize now), and that the war was over, had been for a couple of years. That there was nothing to worry about.

No, nothing to worry about, except that time passed and we moved to California — a much juicier target than the small town — and the threat of nuclear war grew heavy in our hearts. As for nuclear power, well, how could you trust the genie now that it was out of the bottle?

Then, in 1991, I was picked as a participant in a study of people who had been children in eastern Washington when radioactive iodine was release into the air at the Hanford Nuclear Project. That's where the plutonium for the bomb that the United States dropped on Nagasaki, Japan, had been manufactured. After the war, many questions were raised about what went on there and about long-term effects on the people who had worked there and those who lived in the surrounding area, including children born nearby. My name came up when a multi-million-dollar study of folks who lived downwind from Hanford was launched to find out what had happened to kids exposed to the radioactive isotope iodine-131 released by the nuclear project. The study paid for my trip back to Seattle to be tested, and what do you know? I was diagnosed with Hashimoto's thyroiditis and thyroid nodules. The incidence of both conditions was thought to be increased by exposure to radioactive iodine. As it turned out, the incidence wasn't higher than in the general population, but still …[40]

But still, indeed. The accidents at Three Mile Island, Chernobyl, and, more recently, Fukushima in Japan have underlined just how dangerous messing around with the atom can be. So when Gates began talking about the safety of the atom in his book, I was extremely surprised. He asserted that nuclear power was safer than nearly any other kind of power, and that it's unlikely we'll be able to solve our greenhouse gas problems without resorting to a whole lot more of it. He added that he was so convinced of this that he was investing money in a third-generation nuclear reactor scheme.

What? That seemed very off-message if you're talking about avoiding disasters. But then, I had to ask myself, what did I know? Gates has done very well for himself and thousands and thousands of Microsoft investors over the years, and perhaps he was just ahead of the curve on this one. So I began to look around.

A paper published in *Environmental Science and Technology* by Pushker A. Kharecha and James E. Hansen in 2013 is the basis for the claims that nuclear power is extremely safe.[41] Gates refers to it at length, as do Joshua S. Goldstein and Staffan A. Qvist in their book *A Bright Future: How Some Countries Have Solved Climate Change and the Rest Can Follow.*[42]

The basic idea is this: The total number of lives lost directly to nuclear power accidents between 1971 and 2009 is quite small — about 4,900 — but air pollution from other sources of energy, particularly coal, have caused 1.84 million deaths and sixty-four gigatonnes of greenhouse gas emissions. Furthermore, simply switching to more natural gas use "would not mitigate the climate problem and would cause far more deaths than expansion of nuclear power."[43]

Anyone who has seen photos of or experienced first-hand the air pollution caused by coal-powered electrical-generating plants has seen what coal can do to the atmosphere. China, which uses more coal than any other country to produce electricity, has gone so far as to close down manufacturing during winter months because the air above some of its cities becomes unbreathable.[44] Therefore, attributing a massive number of deaths to air pollution from producing electricity with fossil fuels does not seem beyond belief. What is more questionable are the figures for deaths due to nuclear accidents.

Kharecha and Hansen calculate that nuclear accidents account for "approximately 1800 deaths in OECD Europe, 1500 in the United States, 540 in Japan, 460 in Russia (includes all 15 former Soviet Union countries), 40 in China, and 20 in India." They add that a quarter of these deaths "are due to occupational accidents" like scaffolding collapsing, "and about 70% are due to air pollution-related effects (presumably fatal cancers from radiation fallout)."

The incident at Chernobyl provides one of the nastiest statistics. The figures include those from the accident in the spring of 1986, when a somewhat dicey reactor blew up during a routine test gone terribly wrong. Only thirty-one people died immediately, and some involved in heroic — and, thankfully, successful — attempts to keep the accident from becoming even worse lived for decades afterward. For a long time, this figure has been used by proponents of nuclear power generation as proof that even a disaster of this magnitude isn't all that dangerous. Lately, however, more research is suggesting that the toll in human life was greater, although how much greater is a matter of contention. This far out, it is extremely difficult to tease out deaths of old people from the lingering effects of radiation and those from myriad other causes.[45]

As for Three Mile Island, the worst North American accident, no one was killed or injured in the incident, which occurred in 1979. More than thirty years later, however, it appeared that thyroid disease rates in people who were nearby were somewhat higher than expected otherwise.[46]

And then there is Fukushima, where in 2011 a tsunami caused by one of the biggest earthquakes in Japanese history smashed into a reactor on the coast. Some eighteen thousand people were killed in the earthquake and tsunami, and about six thousand people were injured. Among the fatalities are several power plant workers who sustained radiation burns; only one death from the long-term effects of radiation has been recognized, that of a worker who died seven years afterward from lung cancer. The rest were killed in the natural disaster or the evacuation that followed.[47]

All of which adds up to a very small death or injury rate attributed to these three events that have coloured the imagination of millions, even if one takes into account possible delayed effects at Chernobyl. (Note that another disaster was avoided at Chernobyl early in the Ukraine war when the nuclear complex was seized by Russian troops for several weeks. Furthermore, as of late September 2022, fear of accidents seemed to have staved off disaster despite attacks on nuclear facilities elsewhere.)[48]

To approach the question of the safety of nuclear power generation another way, one can compare the number of deaths per quantity of electricity produced. A measure used by the independent website Our World in Data is the number of deaths annually that could be expected from producing a terawatt-hour of electricity, which is the amount used by a town of 150,000 people in Europe in a year. Their calculations suggest that it would take years before anyone would die prematurely if all electricity were produced by nuclear, wind, or solar means, but that twenty-five people a year would die prematurely were the electricity to come from coal-powered generators. Comparable figures are eighteen from oil and three from gas. (Hydro power also is very safe: 1.3 deaths per year per terrawatt hour, a figure that would be even lower if one horrendous dam failure at Banqiao, China, in which 171,000 people died were not factored in.)[49]

Nevertheless, the Fukushima incident shocked the world and spurred several countries to cut back their commitment to nuclear power drastically. Germany, for example, which in 2000 received 29.5 percent of its energy needs from nuclear sources, was phasing out all of its nuclear power installations already, but Fukushima whipped up the process: all were scheduled to be closed by 2022.[50] This withdrawal made coping with embargoes on fossil fuel imports from Russia much harder in the troubled months of the Ukraine war.

Some thirty-three countries operate nuclear power plants, according to *The World Nuclear Industry Status Report 2021*, producing about 10.1 percent of electricity consumed, down from the historic high of 17.5 percent in 1996.[51]

Most countries that are cutting back on nuclear power aim to replace the supply with renewable energy sources like solar and wind, but coal and natural gas will be burned for quite a long time to fill the gap. The result, of course, is more greenhouse gas emissions. Natural gas gives off about half as much CO_2 as coal does per unit of energy produced, but that is still a lot of CO_2. Its impact becomes more problematic when you consider that a good portion of the electricity produced by natural gas doesn't replace electricity from coal but is "extra" electricity. In Germany's case, researchers estimate that the social costs of shutting down nuclear plants amount to twelve billion dollars annually. They note laconically: "Over 70% of this cost comes from the increased mortality risk associated with exposure to the local air pollution emitted when burning fossil fuels. Even the largest estimates of the reduction in the costs associated with nuclear accident risk and waste disposal due to the phase-out are far smaller than 12 billion dollars."[52]

Note also that these calculations were all made before Russia invaded Ukraine and imports of Russian natural gas and oil by European countries became far more problematic. The short-term effect has been higher energy prices in Europe, while the longer-term result may be either a more rapid move toward wind and solar power or, conversely, a slowing of the move away from coal-fired electricity generation.[53]

As for the more distant future, Gates and others argue that new reactor designs will not only produce much safer nuclear generating plants but will also substantially bring down the cost of building the plants and, therefore, the cost of electricity generated. Certainly, China, which is realizing the downside of its considerable economic expansion in terms of pollution, has continued to commission nuclear plants. It currently operates forty-nine nuclear reactors and seventeen more are under construction, and while nuclear power accounts for only 2 percent of the country's electricity, the country clearly intends that it will eventually produce more than any other source.[54]

The United States, by comparison, has ninety-three operational reactors even after shutting down thirty-nine reactors in recent years. They produce about 20 percent of the country's power. Canada has five plants housing twenty-two reactors, which produce about 15 percent of the electricity supply.[55]

All right, so perhaps nuclear power may have a big role to play in fighting climate change, particularly if the new generation of reactors are, as proponents assert, much, much safer than the ones built in the past. But look at Fukushima — is climate change going to be a factor in keeping existing reactors safe?

Goldstein and Qvist point to another nuclear power plant not far from Fukushima that also was affected by that huge earthquake and tsunami in 2011. They say that the reactors near the fishing town of Onagawa came through very well: a fourteen-metre-high seawall prevented major flooding and the reactors shut down "normally and without incident."[56] But at Fukushima, the seawall was less than half that height, and the backup generators were flooded by the massive wave, which was perhaps twelve metres high. The result was that the core reactor overheated into a radioactive mess and released hydrogen gas that exploded, so radioactivity leaked out to the open air and the ocean. The authors hammer home their message: "Note that the problem was not the reactor design but the unforgivable decision to locate all the backup generators together in a location vulnerable to flooding with too small a seawall."

Unforgivable or not, many existing nuclear reactors, as well as ones still on the drawing boards, are located on seacoasts or riverbanks. The reason is simple: the reactor creates heat, which boils water whose steam drives the turbines that make electricity. Somehow or another, that heat must be dissipated, through cooling towers and by passing it through condensers cooled by water from outside. Oceans are particularly good for this sort of heat exchange because the water is almost always considerably colder than the air temperature. This cooling water is shielded from the radioactive parts of the reactor and is discharged a few degrees warmer than it was when it entered the facility.

But that doesn't relieve a certain amount of worry about reactors by the sea, since worldwide one quarter of reactors are on coastlines. In the United States, nine nuclear plants are within three kilometres of the coasts, of which four are particularly vulnerable to storm surges. These include the Turkey Point Station, forty kilometres south of Miami in Florida, which was grazed by Hurricane Andrew in 1992. That it came through as well as it did has been touted as an example of how good its design is. Yet, while in 2019 it got approval to operate until the 2050s, two years later its safety was questioned by the U.S. Nuclear Regulatory Commission because of performance problems.[57] It and its sister plant at Port St. Lucie, Florida, generate about 13 percent of Florida's electricity; natural gas is the state's biggest supplier, around 72 percent of the total.[58]

Hindsight being much better than foresight, it's easy to question why anyone would think some of the sites chosen for nuclear power plants would be safe when faced by rising seas. The answer, basically, is that most of the currently operating reactors were planned long before that threat was on anyone's radar. These include Canada's only nuclear power installation located on tidewater, Point Lepreau in New Brunswick on the Bay of Fundy. Ways must be found to live with the ones that are on the coast, and in the future all decisions on siting must take into consideration rising sea levels.[59] There is growing pressure in the United States and elsewhere, however, to keep existing plants operating and to move on to new, purportedly safer reactors. Indeed, the climate plan drafted for the U.S. Democratic Party in 2019 by John Kerry and Alexandria Ocasio-Cortez advocated creating "'cost-effective pathways' for developing innovative reactors."[60] Other examples of interest include investment by the governments of Canada and New Brunswick in support of developing advanced small modular reactors, at least one of which may eventually be added to the Point Lepreau installation.

In addition, in the summer of 2021, independent research teams reported advances in developing controlled nuclear fusion reactions that augur well for generating power from much safer fusion processes.[61]

Down with Cryptocurrencies

But bringing a lot more nuclear power online will take years. Long before then, we must get serious about combatting climate change. One way is to cut down on energy consumption, which means everything from turning off light bulbs to outlawing cryptocurrencies. Sounds weird, but doing the latter may be one of the big ways to cut back. According to the *New York Times*, cryptocurrency mining now uses about "ninety-one terawatt-hours of electricity annually." That is more than is used by Finland, as much as Washington State uses, "more than a third of what residential cooling in the United States" consumes, and "more than seven times as much electricity as all of Google's global operations."[62] Some of this electricity is generated by renewable means, but every "green" kilowatt used for these financial transactions is one fewer available for other uses. The slack is taken up by electricity generated by conventional means, which means more pollution, more greenhouse gases.

The currency system was designed to cut out the intermediaries by allowing transfer of money from one account to another without banks or governments interfering. One transfer involves millions of calculations in a complicated process that most mere mortals have trouble understanding — and in making these calculations, computers use a lot of electricity. As the cryptocurrency market has exploded, its use of electricity has skyrocketed, too. Since the idea's first appearance in 2008, cryptocurrency has captured a growing share of currency exchanges and now represents billions of dollars. How much varies from moment to moment: on November 8, 2021, one Bitcoin was valued at $84,107.22, but on June 17, 2022, the value had dropped to $26,649.30.[63]

The Chinese government has become increasingly concerned about cryptocurrency for two quite different reasons. First, it would like to control all currency transactions in its domain and has begun moving toward starting its own cryptocurrency system.[64] But also, the enormous amount of electricity needed is a drain on the country's supply, which largely comes from coal-fired generators. In September 2021, because of a suite of problems, including bad air pollution, problems in supplying enough coal for power generation, and a regional drought that cut hydroelectric capacity, a wave of massive power blackouts were ordered. (Note that as mentioned earlier, droughts, in a sort of feast-or-famine scenario, are expected to become more common as weather patterns change.) It is no coincidence that the government began really cracking down on cryptocurrencies at the same time. This would seem to be a no-brainer: the government shifts the electricity supply to the badly stretched commercial and industrial sectors from one that it wants to get rid of. At the same time, it can better control the burning of coal, which is one of its stated goals as it works toward making the country carbon neutral by 2060.[65]

Other governments elsewhere are looking on cryptocurrency with a kinder eye. The mayor of Miami, Francis Suarez, aims to make the city the cryptocurrency capital of the world, despite relatively high electricity prices, while El Salvador has made Bitcoin legal tender and has begun mining it with electricity generated with heat from a volcano.[66] In either case, the process will use electricity that could replace fossil fuel–generated juice, thus reducing CO_2 emissions.

Canadian politician Pierre Poilievre is a great champion of cryptocurrencies. During his campaign for the leadership of the Conservative Party of Canada, he asserted that digital currencies enable the creation of "a new, decentralized, bottom-up economy [that] allow[s] people to take control of their money from bankers and politicians."[67] Just how that would work is not clear. Certainly, the vision he paints in his populist, anti-establishment rhetoric of how cryptocurrencies might be used is completely at odds with serious proposals for using alternate blockchain-generated currencies as a reward for controlling carbon dioxide emissions.[68]

Novelist Kim Stanley Robinson takes that idea seriously and uses it as a key element in his climate-fiction novel *The Ministry for the Future*. In it, a cryptocurrency called carbon coin is used as a lever to get companies, governments, and individuals to switch from fossil fuels to renewable energies and to rewild the world. Sadly, that possibility seems about as likely as a goose laying a golden egg.[69]

Nor is Bill Gates a fan of cryptocurrencies, calling them "100% based on greater fool theory."[70] He also warns people with less money than cryptocurrency fan Elon Musk to stay away from them, although he doesn't mention the energy problems with cryptocurrency in his book.[71] On the other hand, he does outline a number of other high-tech possibilities that might help us control climate change. These include sending very fine particles up into the upper layer of the atmosphere. The aim of this solar engineering would be to reflect energy from the sun, reducing the amount by about 1 percent, which would be enough to lower temperatures around the world by a degree or so.[72] The particles would be sent up in airplanes or rockets; the latter would probably be a more useful outcome of sending things into space than the excursions by three other multi-billionaires in 2021, Richard Branson of Virgin, Jeff Bezos of Amazon, and Musk of Tesla.

Gates calls these geoengineering schemes the equivalent of that glass box with the lever inside that you're supposed to smash open and operate when all else fails. Like the Dutch oceanographer Groeskamp, he may be counting on the wildness of the ideas to capture attention and move public opinion toward vigorous support for climate change abatement. Others, however, are convinced that the idea of solar engineering is worth serious consideration immediately. Near the beginning of *The Ministry for the Future*, Robinson presents a series of responses to a long-lasting heat dome over the Indian subcontinent. These include the Indian government sending up hundreds of flights to inject sulphate into the atmosphere.

This is not an idea to be dismissed as simple science fiction, asserts David Keith, a professor of applied physics and of public policy at Harvard. He believes that even if carbon capture and other methods of cutting back on greenhouse gases work over time, something will be needed to bring down temperatures immediately. He wrote in the *New York Times* at the end of a long hot summer in 2021, "Geoengineering might worsen air pollution or damage the global ozone layer, and it will certainly exacerbate some climate changes, making some regions wetter or drier even as it cools the world. While limited, the science so far suggests that the harms that would result from shaving a degree off global temperatures would be small compared with the benefits."[73]

Elizabeth Kolbert would not agree. In her book *Under a White Sky: The Nature of the Future* she points out that there are so many unknowns with schemes like these that we must insistently ask, "What could possibly go wrong?" She quotes Horace: "Drive out nature though you will with a pitchfork, yet she will always hurry back, and before you know it, will break through your perverse disdain in triumph."[74]

So, What Can We Do to Save Our Civilization?

There are four responses: resist; reconcile and accommodate; retreat and, if necessary, migrate; and — the wild card — remember.

RESIST

Certainly, NEED, the Northern European Enclosure Dam, is the quintessence of resistance. So are all the other seawall projects, of which there are nearly as many as there are seaports. Among them is the six-metre-high seawall across Biscayne Bay that the U.S. Army Corps of Engineers has proposed for Miami, which, wrote the *New York Times*, might be enough "to jolt some Miamians to attention."[75]

But like the strategy of armouring the coast with seawalls and breakwaters, these big projects have major drawbacks, particularly when dealing with floods from increasingly frequent extreme rainfall. What is needed here is to adapt solutions like the Dutch Room for the River program, which showed its merit during those storms that raged through northern Europe in the summer of 2021.

Along the coasts, it's also becoming increasingly clear that hard barriers can be counterproductive. One of the reasons is that when a wave crashes into a seawall and its force is unabated by objects or vegetation, not only is the wall attacked, but the force of the water is directed outward almost as strongly. This means adjacent unprotected properties are walloped even harder. Furthermore, the receding waves can scour the sand away, undermining the wall, and no more sand or sediment will be added to the beach because the wall has stopped erosion for a time. But eventually — inevitably in fact — the wall will give way unless it is constantly maintained, which is a costly affair. Maintenance can run up to fifteen thousand dollars a year for a ten-metre stretch of waterfront, and who is willing to shell that out for decades?

But isn't concrete, that versatile, omnipresent material, nearly permanent? I certainly thought it was a sort of rock of ages until I spent a couple of years researching it for a book. What I discovered is that while some concrete — particularly that made by the Romans a couple of thousand years ago — is still in pretty good shape, most isn't. The formulas used to make modern-day concrete are different, and few people doing construction are planning for a timeline any longer than two or three decades.[76]

But there are other ways to resist the seas.

Mangroves and Other Trees

Bill Gates is bullish on mangroves as a better way of doing that. Certainly, the trees are what stand between the islands of the Sundarbans and obliteration — remember the very name *Sundarbans* means "beautiful forest" in Bangla. Indonesia also, perhaps belatedly, has launched a campaign to rehabilitate its mangroves and to plant more. In late September 2021, President Widodo was everywhere in local media, photographed barefoot and crouching down as he helped children plant seedlings on the island of Riau in the straits between Sumatra and Singapore. By the end of the year, thirty-four thousand hectares of mangrove forests were to be "rehabilitated" as part of a "strategy to defend our national territory, as well as anticipating global climate change," he was reported to say.[77] By 2024 the figure should rise to six hundred thousand hectares. According to the official press release, Indonesia has 20 percent of the world's total mangrove forests, which both absorb carbon dioxide and cut down on coastal erosion.

It should be noted, though, that this mangrove-planting initiative will not include the island of Java, and particularly not Jakarta. While there is a remnant of the once-thick mangrove forest that protected the shore there, it is just a park, a conservancy that was one of the things I'd hoped to see. Of course, I couldn't do that, so my eyes in Jakarta, Alya Zahra Fauzy and Thareq M., took some photos of it for me, as I attempted to get a feel for the city despite Covid-19. For seeing first-hand what a mangrove forest looks like, I have to go back several years to a trip to India and another small park in an urban setting.

The park is on the estuary of the Periyar River, in the Ernakulam district of Kochi, the largest city in India's Kerala state. I was there working on a book about people and Nature in urban settings,[78] and this city, in the best educated, most literate, richest state in the country, had been pointed out to me as good place to explore what can be done in what some call the developing world.

One of the things I had to see, insisted biologist U.K. Gopalan, the former director of India's National Institute of Oceanography, was the mangrove stand in what is now the centre of greater Kochi, which he was instrumental in saving from developers.[79] He was a small, thin man, a vigorous retiree, and he was adamant about the importance of mangroves to the health of the world.

"The mangroves built this coast," he said. The mountains to the east — the Western Ghats, which rise to heights of 2,500 metres — were pushed up by the Earth sometime in the last four million years. Since then, they've been eroding, with sand and silt carried westward by the rivers that course through Kerala. The waves of the Arabian Sea would have carried all this away had it not been for mangroves, Gopalan said. They stabilized the silt and allowed the land to grow slowly, pebble by pebble and grain of sand by grain of sand.

"The mangroves kept the sea from licking away the land," he said. "Without them, the land is disappearing." And he chuckled at the irony of the boulders being trucked down from the mountains to form rip-rap barriers to protect the headlands at nearby Fort Kochi. Silt from these same mountains would do the job better if it were caught in the roots of

Mangroves thrive in brackish water. Here the tide is out.

mangroves, he said. More sombrely, he noted that destruction of the mangrove forests along the coast of Thailand was partly to blame for the severity of damage to shoreline communities during the 2004 tsunami disaster; where there were mangroves to absorb the shock of the great waves, damage was far less.

So I went to check out the small (3.8-hectare) mangrove reserve that Gopalan worked so hard to save. It was midday and very hot when I arrived. There was a sign announcing the reserve at the entrance, but no one else around. The gate stood partway open, as if to say visitors were welcome but not many were expected.

Mangroves thrive in brackish water, which means that the water level of the swamp from which they grow changes with the tides. When I was there, the tide had begun to recede, and I went closer to get a better look. The water did not smell, nor did the mud, although the vegetation along the edges of the water was littered with paper and plastic bags. I even thought that rocks had been deposited by retreating water on some kind of spike-like plant, but a closer look showed the "rocks" to be mud-coated plastic bags.

Something stirred in the undergrowth — a lizard, a snake? I didn't know; I wasn't sure I wanted to know. I thought perhaps I had had enough for one day, even though it was much cooler inside the plot than it was outside. To my North American eyes, the grove held much less charm than a stand of old-growth cedar near the Salish Sea or the green fields of Kamouraska protected by *battures* and *aboiteaux*. But I realize now how wrong I was about its importance.

I'm assured that the trees are still there — Dr. Gopalan's campaign resulted in the refuge being named a bird sanctuary a year or so after I visited. But mangroves are threatened everywhere with repercussions that will have many effects as ocean waters rise and storm surges grow stronger. The Sundarbans are perhaps the greatest example of what is at stake.

Oysters and Other Things

There are other soft techniques that can be used to limit the force of the waves. Among the ones available in more temperate waters is re-establishing oyster beds offshore, which can grow upward faster than the sea level is going to rise.[80] As landscape architect Kate Orff told *The New Yorker* in the spring of 2021, the idea may sound a little bizarre, but it can make a big difference in the dance between sea and shore: "There were oysters, tide pools, grasses, lots of colorful marine life, and they were a big part of New York's coastal-protection system" in the past. These natural defences absorb wave energy and slow the water before it hits the shore. "We've spent the past one hundred years dredging out everything for shipping and hardening the edges," she said. "Now we have a different climate, and we need a different approach."[81]

She is working with the Billion Oyster Project, "a nonprofit that aims to reintroduce the bivalve, in vast quantities, to the waterways of New York City." But this "second nature" approach goes much further: "We have to hit the reset button if we want nature to come back. There's no more natural nature. Now it's a matter of design."

The project at a beach-lined bay off the tip of Brooklyn includes nine breakwater segments that will span the bay. A set of artificial tide pools and rocky basins are intended to be a home to, among other creatures, oysters. The configuration has been extensively tested both by computer simulation and in an Olympic-sized wave pool. Construction began in the summer of 2021, and it's hoped that the artificial reef and oyster beds will dramatically change the pattern of waves, come the next big storm surge.

Certainly, the National Ocean Service of the National Oceanic and Atmospheric Administration says that living shorelines are much more resilient against storms than massive concrete walls. In addition, their research shows that "one square mile of salt marsh stores the carbon equivalent" of nearly 290,000 litres of gas annually, that "marshes trap sediments from tidal waters, allowing them to grow in elevation as sea level rises," and that about five metres of marsh "can absorb 50% of incoming wave energy."[82]

All of these methods can be combined with methods taken from the Netherlands' Room for the River playbook — that is, providing places where rainwater can percolate into the ground or spread out somewhere other than the middle of urban development. The aim is to keep flood waters from arriving at the shoreline just as a storm surge is bearing down on the coast.

RECONCILE AND ACCOMMODATE

But when resisting is not enough or can't be done effectively, we must reconcile ourselves to the situation and find ways to accommodate what is happening. This can involve raising buildings on stilts, as has been done in some places along the coast of the Gulf of Mexico; there are four hundred such buildings on the Texas shore, officially owned by the state and leased to people who love to live on the water — and, frequently, over the water.[83]

In many places where people have lived with periodic flooding for ages, houses on stilts are common, and they are constructed relatively simply. In Bangladesh, for example, bamboo has often been used to build houses on stilts in rural settlements where flooding is frequent; we'll come back to how they're doing it later.[84] That country has also been remarkably successful in reducing the number of deaths during major weather events. In 1970 more than three hundred thousand people died in Cyclone Bhola, but because of early warning systems and shelter construction, the death toll was only three in 2021's Cyclone Yaas, and this despite the fact that anti-Covid-19 measures meant that shelters could not be used to full capacity.[85]

The principal characters in Amitav Ghosh's *The Hungry Tide* find refuge during a particularly strong cyclone in such a storm shelter built by a would-be revolutionary. The man considered the project merely a "social service," and was not sufficiently ambitious to change the world, his wife comments. But Ghosh seems to be saying that it is endeavours like this that will make the difference for people in places like the Sundarbans and elsewhere in the great Ganges-Brahmaputra delta.[86]

Stilt houses in the Florida Keys are considerably more upscale and comfortable than storm shelters like the Bangladeshi ones. Built on concrete pilings anchored into the limestone two metres or more below the waterline, they are designed to withstand hurricanes — and have.[87] Others have gotten onboard after the fact: property owners in New York's Breezy Point on Rockaway Peninsula rebuilt with houses sitting on concrete pillars or high foundations after Hurricane Sandy. Then there is the house literally on Stinson Beach in Marin County north of San Francisco: while surrounding

This house in Stinson Beach, California, really is on the beach.

houses may be raised a metre or so above the high tide line, this one sits on concrete pillars more than four metres high.

This type of construction is very different from the slab-on-grade houses that have been common in parts of North America since the Second World War. Building that sort of housing is relatively cheap: level the site, build a form, pour the concrete several inches thick, and you're good to go. Unless, that is, the house is in a flood-prone area, and then chances of being inundated with increasingly severe weather events are multiplied. Compensating for that danger by raising the house can be expensive. It requires jacking it up off its slab and putting a new foundation underneath, a cost that can run up to $250,000 for a 230 square metre house.[88] And good luck getting flood insurance if you're living in a floodplain. Because of increased claims, insurance is becoming more expensive and harder to get.[89]

Then, of course, there's an idea from the if-you-can't-beat-them department: floating cities, or at least floating buildings. A prototype riverside school in Bangladesh won the Aga Khan Prize for architecture in 2019 for its originality and use of local, sustainable materials and techniques. Located not far from Dhaka, it is made of bamboo and sits on the ground during the dry season. But when the monsoon arrives, it floats on thirty-gallon (114-litre) steel drums encased in bamboo frames.[90]

The Dutch, pioneers in so much related to water and tides, have also experimented with floating buildings. One project, Schoonschip, outside of Amsterdam, is billed as a floating, eco-friendly village. Begun in 2010, the village is not for low-income families, though: prices in 2021 ranged from 300,000 to 800,000 euros.[91] And then there is the luxurious Arkup floating mansion in Miami. With about four hundred square metres of indoor and outdoor space, four bedrooms, and four and a half bathrooms, it is supposed to be ocean-ready but also capable of being hoisted out of the water on retractable legs. In 2020 it was listed for sale at US$5.5 million.[92] Arkup and Waterstudio.NL, a Dutch architectural firm, have collaborated on a series of smaller, modular floating houses that can be put together to form larger units, much like Moshe Safdie's groundbreaking Habitat 67 design, which briefly promised an affordable way to build housing with "for everyone a garden." Arkup and Waterstudio also suggest that their floating structures could be used to create communities, but as with Habitat 67, the costs are going to be far greater than ordinary folk can afford.[93]

All this underlines just how rising sea levels are going to affect millions inequitably. There are worlds of difference between a slum dweller in Dhaka faced with flooding and storm surges and someone whose house overlooking the Salish Sea may slide down a cliff. The first may lose everything her family possesses. The second may have trouble collecting her insurance. Both may decide to move.

Which moves us to the third reaction to rising sea levels: retreat.

RETREAT AND, IF NECESSARY, MIGRATE

As I said back at the beginning, I began to think about the problems presented by higher sea levels a few years ago when we drove along the south coast of the St. Lawrence estuary and saw how, even on a fine day, the waves were crashing within a metre or two of a major regional highway. By chance, the following weekend *Le Devoir*, Quebec's most serious newspaper, carried a special report on how a couple of municipalities in that neighbourhood had begun to persuade people to move away from the shore. Worth looking into, I told myself as I filed the news stories away.

Then came Covid-19 and travel bans, which meant I couldn't see first-hand Indonesia's plan to move its capital from Jakarta. But after mourning the death of that expedition for a while, I realized that taking a first-hand look at the project in Quebec might provide a window on what may be necessary in so many places, big and small, wealthy and poor. Retreat in one form or another is what we're all going to have to do. Indeed, it is what people have started doing all over the world, and what they have done again and again over the long history of rising sea levels.

So in the summer of 2021, when Indonesia was deep into its third wave of Covid-19 and, despite having recently been doubled vaccinated, I was still effectively unable to travel outside Canada, I ventured back to the Bas-Saint-Laurent region, the beautiful borderlands of the St. Lawrence estuary.

In this section, the bottom slopes northward relatively gently from the southern shore. The river itself runs toward the northeast in a channel that is in part defined by the border between two ancient geologic regions. To the south lies the Canadian portion of the Appalachian Mountains, which form the backbone of the eastern United States. Here they curve around and finally end at the tip of the Gaspé Peninsula. A few isolated mini-mountains dot the coastal lowland, relatives of islands that emerge from the estuary's waters. None are very high in contrast to the northern shore, where the relief is much more abrupt. There, the back country is high, and steep escarpments plunge down to the shoreline in many places. This is the edge of the ancient Canadian Shield, made up of some of the oldest rock on the planet. The estuary is much deeper on that side, so the shipping lanes run close to the northern shore.[94]

The village of Sainte-Flavie is on the south shore, with fewer than nine hundred year-round residents.[95] Formerly a small centre that served the farms along the coast and the few fishers who braved the estuary, it has become a tourist hub. The highway that encircles the Gaspé Peninsula begins and ends here, and, perhaps fortuitously, a number of artists and artisans have set up their studios in some of the older buildings. There's a fishing pier too, plus a couple of parks where you can get down to the beach.

Géraldine Colli, the town's *agente de développement*, showed me around the most recently established open spaces. For several years, she'd been helping the owners of properties at risk figure out what do about the threat of rising sea levels and weather extremes. It was clear that she believed in what the town was doing and was proud of how a pilot project aimed at getting people to move out of the way was proceeding.

The day was sunny and warm, which isn't always the case in summer. In fact, many of the places along this stretch of the coast became popular holiday destinations because they were so much cooler in summer than overheated cities like Montreal, and even Boston. In the early twentieth century, before the advent of air conditioning, some wealthy Americans escaped sweltering conditions by visiting the region on their yachts.

On this day, however, a sweater felt good as we walked around the little park where a few new trees had been planted. A village worker was sowing clover seed and setting out bushes chosen because they did well along the coast. Colli explained that the park — where two houses had stood until a couple of years previously — would be developed as another access to the beach because the land had been acquired by the municipality as part of a pilot project to remove buildings from the zone where they were likely to be damaged or destroyed by severe winter storms.

Now, let's be clear about two things. The first is that people around here are not afraid of tough weather. You don't settle in the St. Lawrence valley if you're afraid of winter. From November to March, on average, a total of more than a metre of snow falls here, while the average temperature is well below freezing. Second, the damage that has been done by extreme weather to date is less severe than that experienced in the Sundarbans storm surges or Jakarta's flooded lowlands. It isn't even as bad as what Hurricane Sandy did to the New York and New Jersey coasts in 2012. But this part of Quebec received a wake-up call about what lies ahead on December 6, 2010, with the result that a strategic retreat is being undertaken.

The storm arrived before ice had formed on the estuary. In most winters, the water along the shore would have been frozen solid by then, but in 2010 there was no ice. That meant that the storm's waves were not muffled by frozen water, so they crashed against the shore unimpeded. The rest of the situation recalled the great North Sea storm of 1953, which led to so many projects of resistance in the Netherlands and to the building of the Thames Barrier. This storm brought a tide as high as five metres, the highest recorded in 110 years, strong winds that gusted at near hurricane force just as the tide reached its maximum, and low pressure offshore that whipped up waves enormously. The weather system moved over a vast area, raging from Nova Scotia across the Gaspé Peninsula and into the lower St. Lawrence. No one was killed, but hundreds had to be evacuated and damage was in the hundreds of millions of dollars. Certainly, the toll would have been much higher had the storm-ravaged area been more densely settled.

In Sainte-Flavie, the western part of the little town was hardest hit because of the way the wind was blowing. A half-dozen houses along the shore had to be immediately demolished. There followed considerable reflection on the part of the town, the regional government, and the province. In the aftermath, legislation was passed that allowed no-new-building zones to be established along the coast and regulated what kind of rebuilding could be done within them. At Sainte-Flavie and its neighbour, Sainte-Luce, a research team was hired to analyze each property in a thirty-kilometre stretch and give each a vulnerability index. Then a pilot project was set up through

Two houses at risk from big storms were moved away from the shore in Sainte-Flavie.

which Sainte-Flavie offered the owners of principal residences at risk three options: staying put and taking their chances against high tides and storm surges; moving their houses to a safer location; or abandoning their houses outright and building new ones outside the risk zone. Costs up to $250,000 would be covered, paid from the $5.5 million budget the province put up. Secondary residences were excluded from the deal; while the town welcomes tourists, permanent residents got preference.

As we stood in the park, Colli pointed to two handsomely painted houses; as part of the project, they had been moved across the highway from where we were standing. Immediately to the west of where they had been stood another house, now abandoned, that would shortly be torn down and the land incorporated into the park. Not far away, two new streets abutted the highway. On them several new houses had been built for displaced owners. In all, Colli said with some pride, in the three years the program had been in place, the owners of twenty-two of the thirty properties at risk had agreed voluntarily to move. There were no expropriations, she emphasized, no money spent on court battles over who should leave or who would be responsible for damage when — not if — another storm surge rages out of the estuary. At the end of the exercise, the town put up for auction twenty-one houses that it had acquired during the buyout, and which would otherwise be demolished. Bids for each house started at $3,500, with bidders agreeing to move the house they wanted out of the danger area. The houses were in basically good condition, but moving one would cost the successful bidder up to $150,000. The money raised would be turned back to the provincial government, which had got the project underway with that $5.5 million grant, but the town also hoped to help people who might otherwise be unable to afford a house to acquire one in a time of rapidly increasing housing costs.[96]

In addition, owners along the coast are being encouraged to rethink how they protect their property. If they opt for a bulwark or concrete wall, they are responsible for maintaining it, she said. The original construction can run to fifteen or twenty thousand dollars, with several thousand dollars needed each year for upkeep. Better by far, she suggested, to figure out ways to use softer barriers, like sand dunes and appropriate plantings to protect the shore.

As my husband and I walked along the beach at Sainte-Luce, about fifteen kilometres to the west, I thought about what she had said that afternoon. Because Quebeckers, many of whom travel most years to the beaches of New England, were so eager to get out and about in that second Covid-19 summer, this stretch of the shoreline had masses of visitors and I hadn't been able to find a place in Sainte-Luce to stay. My second choice was a two-storey auberge on the outskirts of Sainte-Luce where we had to climb down a ten-step stairway to get to the beach. I'd chosen the days for our visit because they offered the greatest fluctuation in tides that month, so when the tide was low, we could walk for several kilometres along the Plage de l'Anse aux Coques, Sainte-Luce's locally famous beach.

When we went to bed, the tide was going out. The view from the balcony was pleasant, the sunset nicely coloured. But toward morning I woke up to hear my husband breathing strangely. For a while I lay there, wondering if there was something wrong — until I realized it wasn't him, but the sound of the waves lapping at the base of the cliff because the tide had come in during the night. When we got up, the beach had just about disappeared, with water stretching sixty kilometres to the other side of the estuary. Then my worry turned to what would happen in a storm: the auberge was likely to be in great danger, come the next extreme weather event.

Sainte-Luce has not opted to copy Sainte-Flavie's plan of persuading property owners to move away from the shore, although there are recent additions to the restrictions of what can be built and where in order to minimize dangerous sitings. Given the fact that the houses in Sainte-Luce are much more picturesque and expensive than those in Sainte-Flavie, a $250,000 maximum stipend for moving or abandoning a house at risk wouldn't cover many properties. Rather than doing that, Sainte-Flavie will replenish the beach with a multi-million-dollar project to truck in sand and gravel. Since sandy beaches do help in attenuating the force of waves, this may help. But since this will be the second time in ten years that the town has replenished the beach, I'm willing to take bets on how long the sand will last. Certainly, the force of the waves is going to carry much of that sand back out into the St. Lawrence.

Sea walls and beach at Sainte-Luce on the St. Lawrence estuary.

All right; as I said, this is very small potatoes when compared to the kind of relocation that rising sea levels are going to require. An interesting comparison can be made with what happened along the storm-battered shoreline of New York and New Jersey after Hurricane Sandy. The tropical storm caused US$62 billion worth of damage in the United States, killing at least 125 people there; in the Caribbean, damage was estimated at $315 million, and 160 died.[97] The rebuilding process took several years and involved, as mentioned before, storm-proofing the New York subway system. But for a stretch of shoreline in Staten Island, where housing had been built on a former tidal marsh, authorities opted for buying out the property owners and returning the land to its former state as a barrier marsh.[98]

After the heavy rains that Hurricane Ida brought in 2021, attention has once again returned to housing buyouts, but this time around, voices are being raised asking to include multi-family dwellings as well as single-family ones and to replace existing ones in risky sites with affordable housing in safer places. As Deborah Morris, who formerly led resiliency planning and acquisitions for New York City's Department of Housing Preservation and Development, told *The City*, "The physical footprint of New York may be shrinking because of climate risk, which means you need more housing somewhere else.... The tool of buyouts allows you to help people have another housing option."[99]

Equity concerns have also come up in California, which is toying with the idea of beachfront buyouts. In early summer 2021, both houses of the state's legislature passed a measure setting up a revolving loan fund that would allow coastal municipalities to buy up houses at risk and then rent them out for as long as they were habitable. The rent would go back into the loan fund. No money was attached to the bill, which was one of several designed to address sea-level rise and to manage retreat from the sea. Nevertheless, opponents called it "public housing for the wealthy," charging that legislators were only concerned about people who could afford to own beach property, as less-well-off people in other contexts have gotten little or no support.[100]

Be that as it may, what is certain is that managing a retreat is going to be costly. Post–Hurricane Sandy, New Jersey's buyout program involved $110 million in federal money and bought out 520 properties (on average, $211,538 each), while New York spent $255 million to buy out 654 (on average $389,908). Perhaps it is not surprising that only about forty-five thousand families have agreed to voluntary home buyouts in the United States over the last thirty years.[101]

The high cost of retreat from the coast is one of the reasons why the proponents of NEED, the Northern European Enclosure Dam, say the idea might be cost effective. To retreat is going to cost a lot of money, even if individual property owners are left to pay for their own move and receive nothing for their property while governments pay only for moving infrastructure like sewer lines, waterlines, roads, and electrical service.

Certainly, a growing body of research suggests that retreat must be part of any plan to cope with rising sea levels. Researchers at the University of Delaware insist that a planned retreat must be on the table from the beginning. One of them, A.R. Siders, told *Science Daily* in June 2021, "If the only tools you think about are beach nourishment and building walls, you're limiting what you can do, but if you start adding in the whole toolkit and combining the options in different ways, you can create a much wider range of futures."[102] An article she wrote with Katherine J. Mach points out that managing retreat is also a chance to reconsider and redress historic inequities, since much of the territory at risk is where marginalized and poorer people live. The conflict between Native American tribes and farmers in the Skagit Valley is a case in point, as are neighbourhoods threatened by rising sea levels near Seattle. Solutions to these problems could — if handled generously — redress long-standing inequities.[103]

In order to create a workable plan, Mach and Siders emphasize how useful and important it is, for reasons of efficiency as well as equity, to engage the people who are being affected by rising sea levels. Géraldine Colli in Sainte-Flavie and the Dutch consultants involved in helping Bangladesh plan for the Sundarbans would definitely agree.

When it comes to how much this will cost, a 2011 report by the National Round Table on the Environment and the Economy predicts that, in Canada, flooding due to sea-level rise and storm surges will cause damage costing from one billion to eight billion dollars per year by the 2050s, but over the next century, one billion to six billion dollars could be saved by a combined strategy of not allowing new growth in areas at risk of flooding and gradually abandoning new flood areas.[104]

Of course, there are millions and millions of people who are not going to have any support when they are forced to abandon their homes, or who must cling to what land remains by the flimsiest of constructions. Take the hanging villages of the Sundarbans, for example. One, Kalabogi, received a certain amount of media attention in the summer of 2021, when reporters ventured to investigate how this little settlement, inundated by storm surges in the 2009 Cyclone Aila disaster, has been reduced to a few shacks on stilts. At one time, the village boasted gardens and freshwater ponds, with much of the settlement protected by embankments. But succeeding storms and their raging surges have washed away nearly everything. Travel anywhere now must be done by boat, and the only locally available fresh water is what can be collected during monsoon rains. The houses that remain are raised on stilts a metre or more in height, but most of the young people have left, headed for Dhaka or other centres on higher ground to find work.[105]

Which brings us to the ultimate way to retreat, which is to migrate.

If you've been paying attention all this time, you must have noticed how often I've spoken of people like us moving around. From the very beginning, when we evolved on the plains of Africa, we have gone from place to place, following the seasons and the animals, seeking out the fruits and roots that were ready to harvest, looking for protected coves where we could fish or — after our skill set grew — setting out to sea to try our fortunes. All along we took with us the cultures we had developed, the tools we'd perfected, the songs and stories we'd learned, the customs we'd invented to organize our relations with each other. In that sense, civilizations have endured as people have moved, sometimes because they were forced to, and sometimes because they were going to a place that seemed better. Now we must ask ourselves, Can that continue?

The answer is yes, if we remember, even though our circumstances will change.

So, what must some of us do when — and I fear there is no *if* about it — we fail to control climate change?

Migrate, again! This applies to people living along coastlines everywhere who could be threatened by rising waters next winter or next hurricane season or next monsoon. But the same also goes for people who are suddenly finding summer temperatures impossible, who are threatened by drought and wildfires, who find themselves face to face with worse and worse droughts.

Yes, move, says political scientist and geographer Parag Khanna in his *Move: The Forces Uprooting Us*.[106] He argues that the ultimate answer to the dilemma of climate change is related, perhaps ironically, to the success of our era in reducing infant and maternal mortality, conquering many infectious diseases through immunization and sanitation, and developing effective ways to control births. There are more of us than ever, but predictions suggest that world population will be levelling off soon: families everywhere will cut back on the number of children they have as they come to realize they will not lose most of their offspring in childhood. At the moment, this

210

population decline is occurring mostly in the Global North, which means the population is aging there, while in most places in the Global South, the birth rate is not declining as rapidly as the death rate. This means that populations there continue to grow, and the average age is lower.

Khanna says that large-scale migration to northern countries would solve the dual problems of aging and declining workforces there and of increasing population in places that are going to take huge blows from climate change. Migrants, he asserts, already keep the world economy moving by filling jobs that wouldn't find takers among the privileged of the developed world. Those Bangladeshis selling umbrellas in Venice and sending home remittances are examples, as are those internal migrants who work in Dhaka's garment factories and support their villages in the countryside.

Of course, there will be (and is) opposition to this tactic from "old stock" folks in some of the developed countries, but Khanna notes that "new nationalists" like Donald Trump and Hungary's Viktor Orbán are catering to "an older generation with one foot in the grave." In other words, they won't be around much longer, although in the meantime they can cause a lot of trouble.[107]

Implicit in Khanna's proposal is the need to assimilate the migrants into the receiving societies. This will mean changes, but they will be enriching changes, he would argue. Certainly, to judge from mass migrations of the recent past, the vigour of the resulting mix can produce wonders. Both the United States and Canada, for long stretches of their histories, have been countries of immigrants, for example. These migrations have had their downsides, perhaps the most tragic of which has been the oppression, forced migration, and near extinction of Native American and First Nation populations. Therefore, care must be taken to avoid repeating similar heavy-handed approaches. That is why planning on national and international levels is so important, both in receiving countries and in those who will be losing population.

For those who think this won't work, it's good to remember what happened after the Second World War. In the late 1940s and the 1950s Canada, and to a lesser extent the United States, actively recruited and selected immigrants from postwar Europe to fill expected labour shortages in some industries, like agriculture and mining. The results were positive for all concerned.[108] A faint echo can be seen in the way countries opened their doors to Ukrainian refugees fleeing invading Russians in 2022. The refugees were mostly white and Christian, which made it easier for many of the receiving countries to accept them. But in principle, there is no reason why other migrants couldn't be welcomed warmly.

Managing these migrations will cost money, and this is where the awarding of loss and damage funds to climate-challenged parts of the world, which politicians like Bangladesh's Sheikh Hasina have appealed for, could be very useful. You'll remember she called on COP26 to address the issue "including global sharing of responsibility for climate migrants displaced by sea-level rise, salinity increase, river erosion, floods, and draughts." Ditto those pleas by emissaries from the Maldives and Kiribati whose homes are being rapidly engulfed by rising seas: their aim is to get aid for buttressing their defences against the menace of encroaching waters or to finance wholesale movement of their people elsewhere.[109]

In addition to international migration, internal climate-driven migration is bound to increase. Khanna's colleague and fellow researcher Greg Lindsay decided to leave New York and move to Montreal after Hurricane Sandy because the Canadian destination looked much safer and more interesting. (Khanna says that Canada is as close to a winner as there will be in the climate change sweepstakes.) Some of the rust belt cities of the United States are already looking to have an influx of migrants from hotter, wetter and/or water-starved regions. Duluth, Minnesota, is a good example and was featured in the *New York Times* as one of the most attractive cities around for climate refugees because of its "cold temperatures, abundance of fresh water," and distance from the coast.[110] But, Khanna emphasizes, to make this kind of migration work, we should be planning for it. Duluth has an advantage because it's in a well-off part of the world and has civic leaders eager to capitalize on what the city offers. But planning must go much further than just local initiatives. How to do this is going to be as big a challenge as trying to get countries to agree on a plan to really tackle climate change and to respect their pledges.

REMEMBER!

Our last defence in our struggle to save civilizations is to remember. Again and again, I've trotted out stories from the past to buttress the points I've wanted to make. Some of them are truly ancient; many are more recent; others are from my own experience. All depend on our collective or individual memories.

Education, both formal and informal, is what we need to pass on these and other memories that form the basis of our civilization. (The French, by the way, are well aware of this; *éduquer* means not only to instruct someone, in the sense of what happens in school, but also to bring someone up in a proper manner.)

Ironically, one of the triumphs of this kind of cultural continuity is found in the sad history of the largest forced migration in history, that of the millions of Africans kidnapped, shipped across the Atlantic, and enslaved over four centuries. The odds were against cultural transmission because not only was mortality high on the slave ships and afterward, but also people speaking different languages were brought together specifically so they could not communicate with each other. Nevertheless, they were able to preserve songs, stories, knowledge of medicinal plants, and even religions that live today in what was then called the New World.[111]

But conventional education is also important, and one of the long list of bad things about Covid-19 is the disruption it has caused in the schooling of young people. Take Bangladesh as an example: it has been justly proud of its record in raising literacy and getting its girls into school, but the effort to combat Covid meant that schools were shut down there for 534 days during the pandemic. Efforts to provide distance education through the internet or television during lockdowns were not adequate. Not only has there been a gap in children's schooling, particularly among the least well off, but also many will not be able to come back into the game. Because their families' incomes were drastically cut during the pandemic (despite stipends designed to cushion the shock), many children have already gone to work at whatever little jobs they could find. In order to keep their families' heads above water, both literally and figuratively, many may be forced to keep working.

In addition, many girls who might have continued in school have been married off, in part to reduce the number of mouths to feed in a family but also because in the absence of "progressive" influences, it is easy for folks to slip into old ways like those favouring very early marriage for girls.

Indonesia also was justly proud of how it had reduced poverty and increased educational attainment, pre-pandemic. It, too, attempted to alleviate the economic strain of lockdowns by payments to its citizens, but as elsewhere, considerable ground was lost. How that will play out in the future in each of these cases remains to be seen, but it is clear that an educated, skilled population is important both for the development of the home country and for the perceived value of out-migrants to the countries to which they might go. This is also related in almost a direct line to the amount of money they can send back home.

Another reason for emphasizing the importance of basic education is linked strongly with our hope to keep alive elements of civilization. Remember the case of the Greeks and their writing system mentioned pages and pages back: because of considerable turmoil about 1100 BCE — probably related to some climate disaster — they lost Linear B, their first writing system. The canon of Greek literature was maintained by a powerful oral tradition until a new alphabet, developed by the Phoenicians, those great adventurers and migrants, was introduced. The moral here is that a civilization can survive terrible events if there are ways to safeguard its core stories, as the Greeks did by telling theirs over and over. One corollary is that sometimes outsiders can bring ideas that the receiving society will find wonderfully useful. Another is that education, be it formal or informal, is at the core of civilization.

Amitav Ghosh, who writes so evocatively of the Sundarbans and its climate challenges in *The Hungry Tide* and *Gun Island*, is a case in point. One of the most haunting episodes in the former book is a recounting of the story of Bon Bibi, the protectress of the Beautiful Forest. It's unclear just how old the legend is. Bon Bibi appears to be Muslim, which gives a clue, but because the dangers of the Sundarbans existed long before Islam arrived, it's entirely possible that an old, old tale has been grafted onto newer religious influences. Be that as it may, Bon Bibi is revered by both Hindus and Muslims, and the verse drama in which she rescues an innocent boy is extremely popular in West Bengal and Bangladesh.

A key episode in *The Hungry Tide* tells how an illiterate fisherman can recite verse after verse of the drama which, Ghosh suggests, is a window on how people have lived with the Sundarbans' wonders and dangers for centuries. Passing on the story has been an important element in their culture, in part because it gives lessons on how to live there: take no more than you need, protect the animals and trees, be reverent.[112]

Remember, yes, remember. The worst fate, as Omar El Akkad suggests in his novel *American War*, the very worst thing one can do to someone is to destroy their memories, their stories, their corner of civilization.[113]

PART 5

NO TIME TO WASTE

MUSICAL INTERLUDE

The video opens with cellist Yo-Yo Ma and pianist Kathryn Stott playing their instruments in a window-filled studio looking out on a bay confined by low hills. It's not clear at first where their meditative playing will lead us, just that this is a peaceful place where we can rest and regain strength. Then they glide into the familiar sound of "Over the Rainbow" (by Harold Arlen and Yip Harburg). The song tells of bluebirds flying beyond the rainbow in times of trouble to a place of wonder and dreams coming true. Written at the end of the Great Depression, with the Second World War looming on the horizon, it has become an anthem for all kinds of people seeking a happier, safer place. The video can be found on Yo-Yo Ma's YouTube channel.[1]

Ma and Stott included the song on their album *Songs of Comfort and Hope*, which was an outgrowth of a project Ma started at the beginning of the Covid-19 pandemic. He had begun it almost casually, with homemade videos of himself playing the cello, and then it expanded as he invited others to post videos of the music that was helping them through the difficult time.

"Songs bring a sense of community, identity, and purpose, crossing boundaries and binding us together in thanks, consolation, and encouragement," he and Stott write in the album's liner notes.

Ma, who studied anthropology at Harvard, elaborated on this thought in an interview with the *New York Times*. He told of meeting a group of women in the Kalahari Desert who do trance dancing for hours, and when he asked them why they did it, they said simply, "Because it gives us meaning." That is the reason for culture, he said, which is "everything that humans have invented that helped us survive and thrive. Think about language, think about agriculture, think about navigation, think about engineering. Think about politics."[2]

Think about music, think about stories, I say to myself. Think about how we're going to get through the challenges in front of us now. Of how we're going to save ourselves and civilization this time around.

9

THE RAINBOW

In the run-up to the COP26 meetings in Glasgow, when governments around the world began making pronouncements about what great things they were going to do to fight climate change, Jakarta's administrators announced they were getting ready to forbid groundwater extraction in areas covered by the city-owned piped water service. Only about 65 percent of the megacity's area is hooked up to the system, although plans are to increase the coverage to 100 percent by 2030. For areas covered, however, the new measures would fine industries, hotels, malls, apartment buildings, and offices that pump up groundwater. The step is designed to cut down on subsidence of the land, which compounds the problem of rising seas in the rapidly sinking city.[3] It was good news to the millions of people who will be left behind when Indonesia moves its administrative capital to another island 1,300 kilometres away.

But the larger fate of the people of Jakarta is far from clear. Construction on the massive seawall that is eventually supposed to protect them has progressed slowly for many reasons, not the least the Covid-19 pandemic. When it comes to working out a version of "room for the river" to manage rivers swollen from torrential rains, the current governor is advocating naturalizing the waterways that run through the city, but the peril of storm surges and rising tides will continue as long as sea levels lap higher and higher. And then there is the possibility that a city government less convinced of the effectiveness of that tactic may come to power; the next elections are due to take place in 2024.[4]

Over all looms the very real danger that we're sunk already when it comes to climate change. Before COP26 (the conference's whole name is Conference of the Parties to the UN Framework Convention on Climate Change, which doesn't give much context for the important work it should be doing) began, a series of reports were released that emphasized the precarious state of the world. The pledges that countries made in 2015 at the COP21 meeting held in Paris had not been kept by most of them, and even if they had been, the reports said, global temperatures were rising faster than expected. The rise could be nearer to 3°C instead of the 1.5°C predicted (or hoped for) at that famous meeting.[5] What was needed was for countries to double down on their efforts to reach net zero carbon targets, pledge to get rid of fossil fuels, honour their promises when it came to aid to poorer countries who are bearing the brunt of climate change, and many other things.

That mostly didn't happen. Like a badly loaded barge in troubled waters, coal just about sank the COP26 meetings. India and China, who are the top greenhouse gas emitters, balked at a draft agreement that called for phasing out coal. In the end, the two hundred nations present at the conference agreed to "phasing down" coal use. They also agreed to come to the table at the COP27 meeting with more robust plans to mitigate their carbon footprints — and to give evidence of what they've done to keep the promises they've already made. That last is important: even though the developed nations had pledged collectively to pony up one hundred billion US dollars a year for climate mitigation, the amount collected has fallen far short of that goal. A real loss and damage fund that Bangladeshi prime minister Sheikh Hasina advocates — called reparations in some quarters — would be extremely important to the countries of the world who, through no real fault of their own, are literally drowning as sea levels rise.

When it comes to the attitudes toward coal-fired electricity generation, Bangladesh and Indonesia are a study in contrasts. The former has one coal mine and has proudly announced that it is closing ten coal-fired generating plants. Indonesia, on the other hand, has in recent years been the world's biggest exporter of thermal coal; indeed, said the *Jakarta Post* right after COP26, the country is "addicted" to coal.[6] While it has been making noises that it recognizes the problem of using the substance, there's a disconnect between what the country says and what it does. For example, in November 2021 the Indonesian government announced that it would phase out coal-fired plants by the 2040s and would begin by decommissioning a quarter of its coal capacity by 2030, which is earlier than previous plans. But what wasn't mentioned was that while it was retiring coal plants producing 9.2 gigawatts, it would be building coal plants that would produce 13.8 gigawatts.[7] The new plants may be more efficient and therefore less polluting per unit of power, but they still use coal as fuel. The reason for the difference between Indonesia's plans and those of Bangladesh may partly lie in the fact that, while Jakarta will go under the waves in the future, the government has an exit plan, but Bangladesh's problems are much greater. After all, up to 25 percent of Bangladesh has been flooded during recent cyclones, and that is very hard to ignore.

There was a plethora of other pronouncements coming out of the COP26 meetings: an agreement by one hundred countries including Brazil and Indonesia, both top deforesters, to stop burning and cutting down forests and to encourage reforestation; private sector money for clean technology; and a pact to cut methane emissions by 30 percent by 2030, among other things.[8] Another positive point was the letter signed by U.S. President Joe Biden and Chinese President Xi Jinping agreeing to co-operate on climate matters.

However, when push came to shove — and it seems there was a lot of shoving — there was very little about carbon taxes or tariffs. In Canada, Quebec and British Columbia have programs that put a price on carbon emissions, but in the United States only California, Washington State (beginning in 2023), and eleven states in the northeast (the Regional Greenhouse Gas Initiative) have them. The European Union has been successfully using the mechanism to cut back on carbon emissions and has agreed to levy border taxes (that is, tariffs) on imports from countries that don't tax carbon, thus penalizing manufacturers who move their high-carbon manufacturing offshore. China, however, has begun levying a carbon tax that it says is in effect higher than the ones charged by the EU.[9] Indonesia began imposing a tax on carbon-emitting industries including coal-fired electricity-generating plants in 2022, but plans are not in the works to reduce subsidies for fossil fuels.[10] And collectively the nations at COP26 said, no, they didn't want to get on that bandwagon.[11]

COP26 was too little, too late. A great disappointment. Not at all what climate activists were hoping for. Indeed, doomsayer George Monbiot wrote in the *Guardian* the day after the conference wound up, "After the failure of COP26, there's only one last hope for our survival."[12]

My toes nearly curled when I read that. I'd been brooding about the outcome of the conference, particularly after all the hype that it would be our last chance, but I didn't want to hear my fears expressed in such blunt terms. Yet once I read further, I discovered that, actually, the climate activist has a pretty positive message. What we have to do is get about 25 percent of the population onboard — which is extremely doable, he says — and that is when a tipping point will come into play. What follows will be a cascade of positive feedback that could transform the whole game.[13] Monbiot writes that until now social convention has worked against the movement to combat climate change, but if public opinion can be flipped, convention can "become our greatest source of power, normalising what now seems radical and weird. If we can simultaneously trigger a cascading regime shift in both technology and politics, we might stand a chance. It sounds like a wild hope. But we have no choice. Our survival depends on raising the scale of civil disobedience until we build the greatest mass movement in history, mobilising the 25% who can flip the system. We do not consent to the destruction of life on Earth."

He calls for more and more public action by people concerned about what we are doing to the world, a cry that is likely to be heard in many parts of the developed world. But when it comes to rising powers like India and China, I have my doubts. Just as the Chinese have made sure to encase most of their shoreline in concrete, they are not likely to allow public demonstrations to sway them. Our best hope there is that the very serious consequences of using coal will elicit more mitigating tactics. Don't forget in the last few years China has shut down portions of its industries for certain periods because of air pollution. The country can't afford to allow this to continue. And for what it's worth, China announced in 2021 that it will stop financing coal plants abroad; forty-four had been in the works, for a combined investment of about fifty billion U.S. dollars.

As for India, it aims to produce 450 gigawatts of renewable energy by 2030, but it definitely is continuing to increase its coal-fired capacity. Currently it has 233 gigawatts of coal plants in operation and a further 34.4 gigawatts under construction. Financing for new plants and for increased coal mining is becoming harder, though. According to *Climate Home News*, an independent news site, since the Modi government opened the field to private investment in 2020, not a single foreign investor has expressed interest in financing new coal projects in India.[14]

The 2022 IPCC report on what we must do delivered much the same message about the dangers we face. Unfortunately, the effect of its message was somewhat eclipsed by the fact that it was published just as much of the world's attention was focused on the war in Ukraine. There the dangers are perhaps more understandable. We have fought wars, huge wars, again and again. What we haven't done in a conscious way is change the way of life for several billion people by really getting serious about combatting climate change.

We shall see what happens.

And that takes us back to Covid-19, our response to it, and how that relates to rising sea levels. As I said back at the beginning, the rainbow became the symbol around here for getting through the pandemic: *Ça va bien aller*, it's going to be all right, was the slogan. Of course, it wasn't for many, many folks. The number of deaths worldwide keeps mounting, and the collateral damage in terms of long Covid and disruption of agriculture, trade, manufacturing, the arts, and everything else will not be known for decades. The pandemic has shown people at their best — all those selfless health-care workers, those scientists working flat out to produce a vaccine, those myriad kind gestures by neighbours looking after neighbours. But also, the worst side of many has become increasingly apparent — rich nations grabbing the lion's share of vaccines, some folks wilfully misleading others about the science of the disease and its prevention, enormous selfishness on the part of a huge portion of society.

Given all this, how can we possibly attack the problem of rising sea levels and climate change together? How can we save our civilization? The rainbow story from the Bible isn't much help because, after all, it is a self-centred one. Only a few were saved. God looked out for his people and their animals, not everyone and every beast. As some politicians and military types are fond of saying, hope is not a strategy. We are going to need more than that.

This time around — as has surely been the case over the millennia that we've had to live with rising seas — some people are going to die before their time, with lives turned upside down by storms and storm surges. Many cities will slowly drown, even if they are not severely damaged by specific weather events. Heroic efforts like the dikes and dunes of the Netherlands may make a difference in some regions, but not everywhere. To allay the disaster — and to talk about victory would be false — what is needed is action on a massive scale.

But we must also nurture our stories and songs, because in the past they are what have endured as people have moved and moved and moved again to escape dire circumstances, to seek a better place. True, the songs I've chosen to speak about come from the traditions of Western classical and popular music, but there are so many others that lift the spirit, rend the heart, or set feet dancing. Remember them, remember the stories. Do not simply save them to your computer or the cloud but make them part of your soul.

If nothing else, the Covid-19 pandemic has shown us just how interrelated the world is, how something that occurs in one place can have enormous effects on the other side of the world. Civilization will survive, but it will change. We can only do our best to keep our heads above the water, realizing that what we do will have repercussions for the folks downstream, for the ones also struggling to keep afloat.

At this writing, I haven't taken down the rainbows that my grandkids made, which have been taped to the window of our front door for months and months. When will I? Perhaps one fine day when the pandemic is truly over, but I'm not sure. What's crystal clear, however, is that, unlike Otis Redding in that song that started this inquiry, we must not waste time. The stakes are far too high, and the water is rising.

"It will be all right" was the motto during the Covid-19 pandemic. Let's hope that will be so in the future, too.

ACKNOWLEDGEMENTS

Thanks are due to a number of people: Arif Budiman and his associates in Jakarta, Alya Zahra Fauzy and Thareq M. Back when Arif was a student studying Portuguese in Lisbon, he was asked by the Indonesian Embassy there (why is a long story) to show me around the city. This he did with brio. We've kept in touch over the years, and I've followed with much interest his career as teacher, translator, and amazing polyglot (he speaks something like eleven languages, not counting dialects). When this project was launched, I contacted him to see if he could help me out in Jakarta. His response was an immediate yes, but then, of course, Covid struck. Nevertheless, in the spring of 2021 he arranged for Alya and Thareq to be my eyes in Jakarta. They visited a long list of places I would have visited had I been able to travel, and they took dozens of photographs, some of which are in this book.

Le Conseil des arts et lettres du Québec, who readily agreed to my request to use the travel grant that it accorded me in 2019, meant for my visit to Jakarta, for a number of other expenses (among them books, Zoom, travel to Kamouraska, and photographs) over the following two years to continue the project despite Covid-19.

Kwame Scott Fraser of Dundurn Press for enthusiastically agreeing to back the project on rather short notice, in publishing terms.

Bruce Walsh, formerly publisher at the House of Anansi Press and the University of Regina Press, for adding an *s* to *civilization* when I began talking about the project in 2017. That addition made an enormous difference in how I looked at the whole problem.

Dominic Farrell and Susan Fitzgerald who edited the book for Dundurn. Dominic made excellent suggestions on how to improve it, and Susan found and corrected several embarrassing errors I had made.

My husband, Lee Soderstrom, who has tagged along on many of these adventures without grumbling.

None of these people are responsible for any mistakes I've made, of course.

NOTES

Preface

1   Before the COP26 meetings, the non-profit organization behind the meetings did several simulations of possible climate change trajectories. If we continue with business as usual, we'll end up with 2.7°C warming above pre–Industrial Age levels by 2100. The organization also ran various "optimistic" scenarios taking into account the net zero emissions targets of over 140 countries that have been adopted or are under discussion. Assuming the governments actually make good on their targets, the organization fixes the median warming estimate at 1.8°C, or at least likely to be below 2.0°C. "Temperatures," Climate Action Tracker, accessed July 28, 2022, climateactiontracker .org/global/temperatures. The report of the Intergovernmental Panel on Climate Change was more optimistic in April 2022. It insisted that the 1.5°C benchmark could be maintained if action was taken on numerous fronts: we'll have more to say about this later. Intergovernmental Panel on Climate Change, *Climate Change 2022: Mitigation of Climate Change — Summary for Policymakers*, April 4, 2022, reliefweb.int/report/world/climate -change-2022-mitigation-climate-change-summary-policymakers.

2   Land created by dredging up sediments is frequently called "reclaimed," but that is a misnomer because the land is, in effect, new land. *Reclaimed* can be more accurately used when talking about land surrounded by dikes that keep out river and sea waters so that the ground dries out; given the world's

history of rising sea levels, at one point this land was almost certainly above the tide line. Here *reclaimed* will be used for both sorts.

Chapter 1: The Sinking City

1   View the video at playingforchange.com/videos/sittin-on-the-dock-of-the -bay-song-around-the-world or youtube.com/watch?v=Es3Vsfzdr14.

2   Jonathan Watts, "Indonesia Announces Site of Capital City to Replace Sinking Jakarta," *Guardian*, August 26, 2019, theguardian.com/world/2019/aug/26 /indonesia-new-capital-city-borneo-forests-jakarta.

3   Yulia Savitri, "Two Islands Vanish, Four More May Soon Sink, Walhi Blames Environmental Problems," *Jakarta Post*, January 15, 2020, thejakartapost .com/news/2020/01/15/two-islands-vanish-four-more-may-soon-sink-walhi -blames-environmental-problems.html.

4   "Carbon Footprint by Country 2022," World Population Review, accessed July 28, 2022, worldpopulationreview.com/country-rankings/carbon -footprint-by-country.

5   Hannah Beech et al., "The Covid-19 Riddle: Why Does the Virus Wallop Some Places and Spare Others?," *New York Times* May 3, 2020, nytimes .com/2020/05/03/world/asia/coronavirus-spread-where-why.html.

6   Deden Rukmana, Fikri Zul Fahmi, and Tommy Firman, "Suburbanization in Asia: A Focus in Jakarta," *Indonesia's Urban Studies* (blog), November 17, 2018, indonesiaurbanstudies.blogspot.com/2018/11/suburbanization-in-asia -focus-in-jakarta.html.

7   Tim van Emmerik, "Research: Indonesia's Ciliwung Among the World's Worst Polluted Rivers," *The Conversation*, February 20, 2020, theconversation.com/research-indonesias-ciliwung-among-the-worlds-most -polluted-rivers-131207.

8   Mary Soderstrom, *Road Through Time: The Story of Humanity on the Move* (Regina: University of Regina Press, 2017), 13–17.

9   "Modern Humans Emerged More Than 300,000 Years Ago New Study Suggests," *Science Daily*, September 28, 2017, sciencedaily.com/releases /2017/09/170928142016.htm; Kate Wong, "Ancient Fossils from Morocco Mess Up Modern Human Origins," *Scientific American*, June 8, 2017, scientificamerican.com/article/ancient-fossils-from-morocco-mess-up-modern -human-origins.

10 Jean-Jacques Hublin et al., "New Fossils from Jebel Irhoud, Morocco and the Pan-African Origin of *Homo Sapiens*," *Nature* 546, no. 7657 (2017), nature.com/articles/nature22336; Mary Jackes and David Lubell, "Early and Middle Holocene Environments and Caspian Cultural Change: Evidence

from the Télidjène Basin, Eastern Algeria," *African Archaeological Review* 25, no. 1–2 (2008): 41–55, doi.org/10.1007/s10437-008-9024-2.

11 Israel Hershkovitz et al., "The Earliest Modern Humans Outside Africa," *Science* 359, no. 6374 (2018): 456–59, science.org/doi/10.1126/science.aap8369.

12 Brenna M. Henn, L.L. Cavalli-Sforza, and Marcus W. Feldman, "The Great Human Expansion," *PNAS* 109, no. 44 (October 17, 2021): 17758–64, doi.org/10.1073/pnas.1212380109.

13 The story is complicated by the fact that as ice caps melted and the immense weight of the ice was removed, parts of the Earth's crust rebounded upward. That appears to be true for parts of Scandinavia, the coast of British Columbia, and the Great Lakes basin. For more, see Carol Rasmussen, "Glacial Rebound: The Not So Solid Earth," NASA, August 15, 2015, nasa.gov/feature/goddard/glacial-rebound-the-not-so-solid-earth.

14 For the fascinating story of how the cores were obtained and dated see "Old Ice," in Sarah Dry, *Waters of the World: The Story of the Scientists Who Unraveled the Mysteries of Our Oceans, Atmosphere, and Ice Sheets and Made the Planet Whole* (Chicago: University of Chicago Press, 2019), 230–70.

15 Other factors besides the total amount of liquid water affect the level of the oceans. Warmer water has a larger volume than colder water does, so it will lap higher on the world's shorelines. In addition, the weight of ice caps can weigh down the Earth's surface, but when the ice melts, the surface will rebound, actually thrusting the shoreline higher. More about this in the section on the Salish Sea and Puget Sound.

16 Evelyn Mervine, "Geology Word of the Week: E is for Eustasy," *Georneys* (blog), December 1, 2010, blogs.agu.org/georneys/2010/12/01/geology-word-of-the-week-e-is-for-eustasy. *Eustatic* was borrowed into English from German *eustatische*, which was coined from Greek *eu* (well) + *statikos* (static, standing).

17 Fossils of *Homo floresiensis*, dubbed "the Hobbit," have been found on Flores too. Their history remains a puzzle, although one school of thought is that they were somehow related to *Homo erectus*. "*Homo floresiensis*," National Museum of Natural History, July 1, 2022, humanorigins.si.edu/evidence/human-fossils/species/homo-floresiensis.

18 Robert G. Bednarik, "The Earliest Evidence of Ocean Navigation," *International Journal of Nautical Archaeology* 26, no. 3 (August 1997): 183–91, doi.org/10.1111/j.1095-9270.1997.tb01331.x.

19 Donald Lee Johnson, "Problems in the Land Vertebrate Zoogeography of Certain Islands and the Swimming Powers of Elephants," *Journal of Biogeography* 7, no. 4 (December 1980): 383–98, doi.org/10.2307/2844657; "Can Elephants Swim?," Wild Animal Park, April 3, 2014, wildanimalpark.org/can-elephants-swim.

20 There are tantalizing suggestions that we do, however. See Adam P. Van Arsdale, "*Homo Erectus* — A Bigger, Smarter, Faster Hominin Lineage," *Nature Education Knowledge* 4, no. 1 (2013): 2, nature.com/scitable /knowledge/library/homo-erectus-a-bigger-smarter-97879043; Alan R. Rogers et al., "Neanderthal-Denisovan Ancestors Interbred with a Distantly Related Hominin," *Science Advances* 6, no. 8 (February 21, 2020), science .org/doi/10.1126/sciadv.aay5483; and Ann Gibbons, "Mysterious 'Ghost' Populations Had Multiple Trysts with Human Ancestors," *Science*, February 20, 2020, doi.org/10.1126/science.abb3777.

21 K. Westaway et al., "An Early Modern Human Presence in Sumatra 73,000–63,000 years ago," *Nature* 548 (2017): 322–25, doi.org/10.1038 /nature23452.

22 K. Norman et al., "An Early Colonisation Pathway into Northwest Australia 70–60,000 Years Ago," *Quaternary Science Reviews* 180 (January 15, 2018): 229–39, dx.doi.org/10.1016/j.quascirev.2017.11.023.

23 Peter Raby, *Alfred Russel Wallace: A Life* (Princeton: Princeton University Press, 2001).

24 What to call the archipelago that includes all of the islands between Southeast Asia and Australia is a question I wrestled with for months. There are an estimated twenty-five thousand islands in the area, divided among several present-day countries, the biggest of which are Indonesia, the Philippines, and Malaysia. Wallace called them the Malaysia archipelago, while colonial powers called them the Spice Islands and the East Indies. Maritime Southeast Asia is also used. But none are satisfactory, to my mind. Hence the Ocean Isles, which doesn't have colonial echoes or cause confusion with the various current political arrangements.

25 *The Forgotten Voyage: Alfred Russel Wallace and His Discovery of Evolution by Natural Selection*, directed by Peter Crawford (BBC, 1983), part of The World About Us series, youtube.com/watch?v=Z1eQ6DadodA.

26 Specifically, James Ussher, a seventeenth-century Irish archbishop, calculated that creation began on Saturday, October 22, 4004 BCE, at 6:00 p.m. His basis was the ages of Abraham's family. He apparently didn't make any adjustment for the fact that God would have wanted to rest on the Sabbath. "Can We Measure the Earth's Age According to the Bible?," *Huffpost*, February 14, 2017, huffpost.com/entry/can-we-measure-the-earths_b _14748994.

27 "Indian Ocean Tsunami: Then and Now," *BBC News*, December 25, 2014, bbc.com/news/world-asia-30034501.

28 Alex Fox, "Indonesian Volcano 'Anak Krakatau' Fired Lava and Ash Into the Sky Last Weekend," *Smithsonian Magazine*, April 14, 2020, smithsonianmag .com/smart-news/indonesian-volcano-anak-krakatau-fires-lava-and-ash-sky

-180974665 and David Bressan, "The Eruption of Krakatoa Was the First Global Catastrophe," *Forbes*, August 31, 2016, forbes.com/sites/davidbressan/2016/08/31/the-eruption-of-krakatoa-was-the-first-global-catastrophe/?sh=7019a71d2f1e.

29 Tara Parker-Pope, "The Human Body Is Built for Distance," *New York Times*, January 27, 2009, nytimes.com/2009/10/27/health/27well.html.

30 For more about this idea, see the chapters "Bottleneck on the Road from Eden" (pages 11–25) and "The War Against the Forest" (pages 37–41) in *Road Through Time: The Story of Humanity on the Move*. See also Gordon H. Orians and Judith H. Heerwagen, "Evolved Responses to Landscapes," in *The Adapted Mind: Evolutionary Psychology and the Generation of Culture*, ed. Jerome H. Barkow, Leda Cosmides, and John Tooby (Oxford: Oxford University Press, 1992), 555–79.

31 Norman, "An Early Colonisation Pathway."

32 Sue O'Connor, Rintaro Ono, and Chris Clarkson, "Pelagic Fishing at 42,000 Years Before the Present and the Maritime Skills of Modern Humans," *Science* 334, no. 6059 (2011): 1117–21, doi.org/10.1126/science.1207703.

33 Mark Arnoff and Janie Rees-Miller, eds., *The Handbook of Linguistics* (Hoboken, NJ: Wiley Blackwell, 2003), 705. In Tanzania, Swahili was chosen as the national language for much the same reason.

34 Peter Bellwood, "Austronesian Prehistory in Southeast Asia: Homeland, Expansion and Transformation," in *The Austronesians: Historical and Comparative Perspective*, ed. Peter Bellwood, James J. Fox, and Darrell Tryon (Canberra, Australia: ANU Press, 2006), 103–18, jstor.com/stable/j.ctt2jbjx1.8.

35 M. Aubert et al., "Pleistocene Cave Art from Sulawesi, Indonesia," *Nature* 514 (2014): 223–27, doi.org/10.1038/nature13422.

36 M. Sytayasa, "Notes on the Buni Pottery Complex, Northwest Java," *Mankind* 8, no. 3 (June 1972): 182–84, doi.org/10.1111/j.1835-9310.1972.tb00433.x.

37 See "Photos: National Museum Part 2 — The Gedung Arca," *Queen of the East* (blog), November 4, 2013, queenoftheeast.wordpress.com/tag/inscription/; J. Noorduyn and H.Th. Verstappen, "Purnavarman's River-Works near Tugu," *Bijdragen tot de Taal-, Land- en Volkenkunde (Journal of the Humanities and Social Sciences of Southeast Asia)* 128, no. 2/3 (1972): 298–307.

38 Julie Romain, "Indian Architecture in the 'Sanskrit Cosmopolis': The Temples of the Dieng Plateau," in *Early Interactions Between South and Southeast Asia: Reflections on Cross-Cultural Exchange*, ed. Pierre-Yves Manguin, A. Mani, and Geoff Wade (Cambridge: Cambridge University Press, 2015), 302–16.

39 "Prambanan Temple Compounds," UNESCO World Heritage Convention, accessed July 31, 2022, whc.unesco.org/en/list/642.

40   S. Supomo, "Indic Transformation: The Sanskritization of *Jawa* and the Javanization of the *Bharata*," in *The Austronesians: Historical and Comparative Perspectives*, ed. Peter Bellwood, James J. Fox, and Darrell Tryon (Canberra, Australia: ANU Press, 2006), 308–32, doi.org/10.22459/A.09.2006.15.

41   Sir Thomas Stamford Raffles, *The History of Java*, vol. 2, 2nd ed. (London: John Murray, 1830), 70, accessed at gutenberg.org/ebooks/49843.

42   "Share of Indonesian Population in 2010, by Religion," Statista, statista.com /statistics/1113891/indonesia-share-of-population-by-religion.

43   I found a photo of the *padrão* in one of the books I acquired years before I started this project. It was stuck away, almost forgotten, in the travel section of my bookcases with my other books about Portugal. Right next to it was another book I picked up on one of my visits to Lisbon, *A Influência Portuguesaa na Indonésia* by António Pinto de França. Both came my way through the good offices of the Indonesian Embassy to Portugal. My guide to Lisbon in 2009 when I was researching my book *Making Waves: The Continuing Portuguese Adventure* was Arif Budiman, a young Indonesian man who was studying Portuguese there and also teaching Indonesian at the Indonesian Embassy. Over the years I've followed Arif's career, and for this project, he arranged for some of his students to be my eyes in Jakarta when it became clear I wouldn't be able to visit myself. You never know when paths will cross again!

44   "East Timor," *Britannica*, last modified July 11, 2022, britannica.com/place /East-Timor.

45   Tomé Pires, Portuguese explorer and chronicler, notes that the Javanese didn't have either butter or cheese at the time he visited, around 1520. *The Suma Oriental of Tomé Pires*, trans. Armando Cortesão (London: Hakluyt Society, 1944), 181. The Portuguese introduced dairy products and how to make them, so it is not surprising that the Bahasa Indonesia words for *butter* and *cheese* are *mentega* and *keju*, very much like the Portuguese words *manteiga* and *queijo*. The respective Dutch words are *boter* and *kaas*.

46   Denys Lombard, *Le carrefour javanais: Essai d'histoire globale* (Paris: Édition de l'école des hautes études en sciences sociales, 2004), 62–68.

47   John Crawfurd, *Descriptive Dictionary of the Indian Islands and Adjacent Countries* (London: Bradbury & Evans, 1856), 45, archive.org/stream/ldpd _6769878_000/ldpd_6769878_000_djvu.txt.

Chapter 2: Those Mythic Floods

1   "Weather of Jakarta," *Indonesia Point*, accessed July 31, 2022, indonesiapoint.com /tourist-attractions/jakarta/weather-of-jakarta.html.

2   "BMKG Warns of Floods as Rainy Season Arrives," *Jakarta Post*, September 23, 2020, thejakartapost.com/paper/2020/09/22/bmkg-warns-of-floods-as-rainy-season-arrives.html.

3   Ray P. Norris and Barnaby R.M. Norris, "Why Are There Seven Sisters?," preprint, submitted December 18, 2020, 6, arxiv.org/abs/2101.09170.

4   Ker Than, "Noah's Ark Found in Turkey?," *National Geographic*, April 30, 2010, nationalgeographic.com/news/2010/4/100428-noahs-ark-found-in-turkey-science-religion-culture.

5   Cristian Violatti, "Greek Alphabet," *World History Encyclopedia*, February 5, 2015, ancient.eu/Greek_Alphabet.

6   Benjamin Jowett, trans., *Plato's Phaedrus*, Sections 274b–278e, socrates.acadiau.ca/courses/engl/rcunningham/1103/Phaedrus.pdf.

7   Patrick Nunn, "Australian Aboriginal Traditions About Coastal Change Reconciled with Postglacial Sea-Level History: A First Synthesis," *Environment and History* 22, no. 3 (August 2016): 397, doi.org/10.3197/096734016X14661540219311.

8   George Fletcher Moore, *A Descriptive Vocabulary of the Language in Common Use Amongst the Aborigines of Western Australia* (London: Wm. S. Orr, 1842), as cited in Patrick Nunn, *The Edge of Memory: Ancient Stories, Oral Tradition and the Post-Glacial World* (London: Bloomsbury Sigma, 2018).

9   E.R. Gribble, *The Vanishing Aboriginals of Australia* (Sydney: Australian Board of Missions, 1933), as cited in Nunn, "Australian Aboriginal Traditions," 414.

10 Nunn, "Australian Aboriginal Traditions"; see also Nicholas Reid, Patrick Nunn, and Margaret Sharpe, "Indigenous Australian Stories and Sea-Level Change," in *Indigenous Languages: Their Value to the Community: Proceedings of the 18th FEL Conference* (Bath, England: Foundation for Endangered Languages, 2013): 82–87.

11 Reid, Nunn, and Sharpe, "Indigenous Australian Stories," 86.

12 Qinglong Wu et al., "Outburst Flood at 1920 BCE Supports Historicity of China's Great Flood and the Xia Dynasty," *Science* 353, no. 6299 (August 5, 2016): 579–82, doi.org/10.1126/science.aaf0842.

13 "Crater Lake: Rich in History, Resources & Adventure," National Park Reservations, accessed July 31, 2022, nationalparkreservations.com/article/craterlake-national-park-history-adventure (page discontinued).

14 Lisa-Marie Shillito et al., "Pre-Clovis Occupation of the Americas Identified by Human Fecal Biomarkers in Coprolites from Paisley Caves, Oregon," *Bulletin Science Advances* 6, no. 29 (July 15, 2020), advances.sciencemag.org/content/6/29/eaba6404.

15 T.S. Subramanian, "The Secret of the Seven Pagodas," *Hindu*, May 20, 2005, frontline.thehindu.com/other/article30204754.ece.

16 Anastasia G. Yanchilina et al., "Compilation of Geophysical, Geochronological, and Geochemical Evidence Indicates a Rapid Mediterranean-Derived Submergence of the Black Sea's Shelf and Subsequent Substantial Salinification in the Early Holocene," *Marine Geology* 383 (January 1, 2017): 14–34, doi.org /10.1016/j.margeo.2016.11.001.

17 Quirin Schiermeier, "River Reveals Chilling Tracks of Ancient Flood: Water from Melting Ice Sheet Took Unexpected Route to the Ocean," *Nature* 464, no. 657 (2010), nature.com/articles/464657a.

18 Jens O. Herrle et al., "Black Sea Outflow Response to Holocene Meltwater Events," *Scientific Reports* 8 (2018): 4081, nature.com/articles/s41598-018-22453-z.

19 Gen. 9:15–17.

20 Vince Gaffney, Simon Fitch, and David Smith, *Europe's Lost World: The Rediscovery of Doggerland*, Research Report no. 160 (London: Council for British Archaeology, 2009), researchgate.net/publication /259639459_Europe%27s_Lost_World_The_Rediscovery_of_Doggerland. See also British writer Pete Kelly's very interesting documentary, *How Doggerland Sank Beneath the Waves (500,000–4000 BC)*, youtube.com /watch?v=DECwfQQqRzo.

21 As cited in Matthew Hatvany, *Marshlands: Four Centuries of Environmental Change on the Shores of the St. Lawrence* (Quebec City: Les Presses de la Université Laval, 2004), 29.

22 "History of Malaria," Rentokil PCI, accessed July 31, 2022, rentokil -pestcontrolindia.com/vector-control/malaria/malaria-history and Institute of Medicine (US) Committee on the Economics of Antimalarial Drugs, *Saving Lives, Buying Time* (Washington, D.C.: National Academies Press, 2004), ncbi.nlm.nih.gov/books/NBK215638.

23 Gaffney et al., *Europe's Lost World*, 21.

24 As cited in Gaffney et al., *Europe's Lost World*, 28.

25 Pål Nymoen and Birgitte Skar, "The Unappreciated Cultural Landscape: Indications of Submerged Mesolithic Settlement Along the Norwegian Southern Coast," in *Submerged Prehistory*, ed. Jonathan Benjamin, Clive Bonsall, Catriona Pickard, and Anders Fischer (Oxford: Oxbow Books, 2011), 40.

26 Bernhard Weninger et al., "The Catastrophic Final Flooding of Doggerland by the Storegga Slide Tsunami," *Documenta Praehistorica* 35 (January 2008): 16, doi.org/10.4312/dp.35.1.

27 Marcel J.L.Th. Niekus et. al., "'Peopling Doggerland': Submerged Stone Age Finds from the Dutch North Sea," Rijksmuseum van Oudheder Rotterdam, academia.edu/34647217/Peopling_Doggerland_Submerged_Stone_Age _finds_from_the_Dutch_North_Sea.

Chapter 3: Fighting Back

1 View the video at youtube.com/watch?v=AGz5Yi75eKU.

2 Ehud Galili et al., "A Submerged 7000-Year-Old Village and Seawall Demonstrate Earliest Known Coastal Defence Against Sea-Level Rise," *PLOS* (December 18, 2019), doi.org/10.1371/journal.pone.0222560.

3 Patrick Nunn, "In Anticipation of Extirpation: How Ancient Peoples Rationalized and Responded to Postglacial Sea Level Rise," *Environmental Humanities* 12, no. 1 (May 2020): 119–20, doi.org/10.1215/22011919-8142231.

4 Nunn does not include the story of the lost city of Atlantis among them because it was made up by Plato to prove a point about morality. Willie Drye, "Atlantis," *National Geographic*, nationalgeographic.com/history/article/atlantis.

5 "Two Wonder Boys Who Controlled Floods," *Tamil and Vedas* (blog), February 18, 2013, tamilandvedas.com/2013/02/18/two-wonder-boys-who -controlled-floods.

6 "Mill Network at Kinderdijk-Elshout," UNESCO World Heritage Convention, accessed July 31, 2022, whc.unesco.org/en/list/818.

7 Marinus Polak and Laura Koolstra, "A Sustainable Frontier? The Establishment of the Roman Frontier in the Rhine Delta," *Jahrbuch des Römisch-Germanischen Zentralmuseums* 60 (2013): 391.

8 Anton Ervynck et al., "Human Occupation Because of a Regression, or the Cause of a Transgression? A Critical Review of the Interaction Between Geological Events and Human Occupation in the Belgian Coastal Plain During the First Millennium AD," *Probleme der Küstenforschung im südlichen Nordseegebiet* 26 (1999): 96–121.

9 Karl-Ernst Behre, "A New Holocene Sea-Level Curve for the Southern North Sea," *Boreas* (January 2007), doi.org/10.1111/j.1502-3885.2007.tb01183.x.

10 Caius Cornelius Tacitus, *Tacitus: The Histories* 4.3.12, vol. 2, trans. W. Hamilton Fyfe (London: Clarendon Press 1912), accessed at gutenberg.org /files/16927/16927-h/16927-h.htm.

11 Frank van Schoubroeck and Harm Kool, "The Remarkable History of Polder Systems in the Netherlands" (paper presented at the International Consultation on Agricultural Heritage Systems of the 21st Century, Chennai, India, February 18, 2010), fao.org/fileadmin/templates/giahs/PDF/Dutch -Polder-System_2010.pdf.

12 L.M. Bouwer and P. Vellinga, "On the Flood Risk in the Netherlands," in *Flood Risk Management in Europe*, ed. Selina Begum, Marcel J.F. Stive, and Jim W. Hall (New York: Springer, 2007): 469–84.

13 Ulf Büntgen et al., "Cooling and Societal Change During the Late Antique Little Ice Age from 536 to Around 660 AD," *Nature Geoscience* 9, no. 3 (February 2016), doi.org/10.1038/ngeo2652.

14 Sarah Zielinski, "Sixth-Century Misery Tied to Not One, But Two, Volcanic Eruptions," *Smithsonian Magazine*, July 8, 2015, smithsonianmag.com/science-nature/sixth-century-misery-tied-not-one-two-volcanic-eruptions-180955858.

15 Julia Pongratz et al., "Coupled Climate–Carbon Simulations Indicate Minor Global Effects of Wars and Epidemics on Atmospheric CO2 Between AD 800 and 1850," Carnegie Science Global Ecology Labs, accessed July 31, 2022, www-legacy.dge.carnegiescience.edu/labs/caldeiralab/Caldeira_research/Pongratz_CoupledClimate.html. This is a complicated and controversial idea. It is clear that vegetation can serve as a sink for carbon, which is the reason why tree planting and ending deforestation are seen as essential to combatting climate change. For more, see the section "Trees and Other Carbon Sinks" in chapter 8.

16 Brian Fagan, *Floods, Famines and Emperors: El Nino and the Fate of Civilizations* (New York: Basic Books, 1999), 194.

17 D.F.A.M. van den Biggel et al., "Storms in a Lagoon: Flooding History During the Last 1200 Years Derived from Geological and Historical Archives of Schokland," *Netherlands Journal of Geosciences* 93, no. 4 (2014): 175–96, doi.org/10.1017/njg.2014.14.

18 For a delightful account of the founding of the university and the botanical garden, where Linnaeus developed his taxonomic system for categorizing plants, see *Hortus Academicus Lugduno-Batavus, 1587–1937* by H. Veendorp and L.G.M. Baas Becking (Leiden: Rijksherbarium/Hortus Botanicus, 1990). I have seen others dismiss the story of the choice given the good folks of Leiden, but, hey, it's a really good one, and these scholars vouch for it.

19 J.S. Marshall Radar Observatory, "The Stormy Weather Group," McGill University, archived July 6, 2011, web.archive.org/web/20110706185139/radar.mcgill.ca/who-we-are/history.html.

20 Megan Garber, "Dan Rather Showed the First Radar Image of a Hurricane on TV," *The Atlantic*, October 29, 2012, theatlantic.com/technology/archive/2012/10/dan-rather-showed-the-first-radar-image-of-a-hurricane-on-tv/264246.

21 A. Muh. Ibnu Aqil, "Floods Cause Widespread Disruption in Jakarta," *Jakarta Post*, February 22, 2021, thejakartapost.com/paper/2021/02/21/floods-cause-widespread-disruption-in-jakarta.html.

22 "Monthly Weather — Manggarai, Jakarta, Indonesia," Weather Channel, weather.com/en-CA/weather/monthly/l/a7eec0ec042c8dc5ba2e285843ba334c659114617043718b7f897bef2c22b1ec.

23 Vela Andapita and Sausan Atika, "'It Was Scary': Wall Collapse Raises Concerns About Coastal Safety in Jakarta," *Jakarta Post*, December 7, 2019, thejakartapost.com/news/2019/12/07/it-was-scary-wall-collapse-raises-concerns-about-coastal-safety-in-jakarta.html.

24 Budi Sutrisno, "Poor River Restoration Blamed for Jakarta's Floods," *Jakarta Post*, February 26, 2021, thejakartapost.com/paper/2021/02/25/poor-river -restoration-blamed-for-jakartas-floods.html.

25 "War For Water: What Happens When Asia's Rivers Dry Up?," directed by Leo Gizzi, *The Longest Day*, uploaded September 19, 2020, on CNA Insider, youtube.com/watch?v=oL-ejcX7GLA.

26 CDPQ, "DP World and CDPQ Sign Long-Term Port and Logistics Park Agreement with Maspion Group in Indonesia," news release, March 5, 2021, cdpq .com/en/news/pressreleases/dp-world-and-cdpq-sign-long-term-port-and -logistics-park-agreement-with-maspion.

27 Pires, *The Suma Oriental*, 192–93.

28 ASEAN Coordinating Centre for Humanitarian Assistance, "Indonesia, Flooding in Gresik Regency, East Java (12:00 Mar 17 2021)," news release, March 17, 2021, reliefweb.int/report/indonesia/indonesia-flooding -gresik-regency-east-java-1200-mar-17-2021.

29 Marchio Irfan Gorbiano and Budi Sutrisno, "'Mudik' Banned — Again," *Jakarta Post*, March 26, 2021, thejakartapost.com/news/2021/03/26/mudik -banned-again.html.

Chapter 4: Wresting a Home from the Sea

1 J. Sherman Bleakney, *Sods, Soil, and Spades: The Acadians at Grand Pré and Their Dykeland Legacy* (Montreal and Kingston: McGill-Queen's University Press, 2004).

2 Bleakney, 37.

3 Kimberly R. Sebold, *From Marsh to Farm: The Landscape Transformation of Coastal New Jersey* (Washington, D.C.: National Park Service, 1992), 5.

4 M.J. Harvey, "Salt Marshes of the Maritimes," *Nature Canada* 2, no. 2 (1973): 22–26.

5 One possible origin is Algonquin *akamaraska* (*akân*, edge of the water, and *ayashaw*, rushes). Another is the Mi'kmaq words *kamoo* (expanse) and *askw* (rushes). Both accurately describe the landscape. Gouvernment du Québec, Commission de toponymie, "Kamouraska," accessed July 31, 2022, toponymie.gouv.qc.ca/ct/ToposWeb/fiche.aspx?no_seq =186980.

6 Lao Tzu, *The Sayings of Lao-Tzu*, trans. Lionel Giles (London: Hazell, Watson & Viney, 1905), 46, accessed at sacred-texts.com/tao/salt/salt01.htm.

Chapter 5: Shanghai

1 View the video at youtube.com/watch?v=ItAxKVP3jlU.

2 Betty Peh-T'i Wei, *Shanghai: Crucible of Modern China* (Hong Kong: Oxford University Press, 1990), 5.

3 Henry Shum, "China Eco-City Tracker: Coming Clean on Shanghai's Water Worries," Asia Pacific Foundation of Canada, March 7, 2019, asiapacific.ca /blog/china-eco-city-tracker-coming-clean-shanghais-water-worries; "Fresh Water for Shanghai," NASA Earth Observatory, accessed July 31, 2022, earthobservatory.nasa.gov/images/89996/fresh-water-for-shanghai.

4 Catherine Seavitt, "Yangtze River Delta Project," in "Rethinking Infrastructure," ed. Stephanie Carlisle and Nicholas Pevzner, special issue, *Scenario*, no. 3 (Spring 2013), scenariojournal.com/article/yangtze-river-delta-project.

5 To cut down on silting, in the early fifteenth century the Song River (now called Suzhou Creek) was dredged and rerouted to join the Huangpu. It previously had run eastward from Lake Tai and the silk town of Suzhou to empty beyond Shanghai in the sea. When the portion east of Shanghai was cut off, adding more water to the Huangpu, which emptied directly into the Yangtze, silt was carried northward and into the main flow of the larger river, where it was less likely to be washed back at high tide. See Linda Cooke Johnson, *Shanghai: From Market Town to Treaty Port, 1074–1858* (Palo Alto, CA: Stanford University Press, 1995), 154.

6 Johnson, *Shanghai*, 78.

7 Dhritiraj Senguptaa, Ruishan Chena, and Michael E. Meadows, "Building Beyond Land: An Overview of Coastal Land Reclamation in 16 Global Megacities," *Applied Geography* 90 (December 5, 2017): 229–38 , doi.org/10 .1016/j.apgeog.2017.12.015.

8 Jeremy Page, "1989 and the Birth of State Capitalism in China," *Wall Street Journal*, May 31, 2019, wsj.com/articles/1989-and-the-birth-of -state-capitalism-in-china-11559313717.

9 In my book *Concrete: From Ancient Origins to a Problematic Future* (Regina: University of Regina Press, 2020), I talk a lot about how this extremely ambitious Chinese effort led to the country using more concrete between 2010 and 2013 than the United States used in the twentieth century. For more about the new cities themselves, see Wade Shepard's *Ghost Cities of China: The Story of Cities Without People in the World's Most Populated Country* (London: Zed Books, 2015).

10 "Lingang New City," GMP, world-architects.com/en/gmp-architekten -von-gerkan-marg-und-partner-hamburg/project/lingang-new-city?nonav=1.

11 Yang Jian, "Lingang Sponge City Nears Completion," *Shine News*, August 22, 2019, shine.cn/news/metro/1908220642.

12 Mia Lu and Joanna Lewis, *China and US Case Studies: Preparing for Climate Change — Shanghai: Targeting Flood Management*, Georgetown Climate

Center, August 2015, georgetownclimate.org/files/report/GCC-Shanghai _Flooding-August2015.pdf.

13   Zhang Cao, "Govt to Reclaim Land from Sea Despite Environment Concerns," *Global Times*, March 11, 2010, globaltimes.cn/content/511991.shtml.

14   Coco Liu, "Shanghai Struggles to Save Itself from the Sea," *Scientific American*, September 27, 2011, scientificamerican.com/article/shanghai-struggles -to-save-itself-from-east-china-sea.

15   There have been a series of studies since 2010 that have forecast problems for many cities on the seas, including several that class Shanghai as one of the world's most endangered cities. See Jie Yin et al., "Multiple Scenario Analyses Forecasting the Confounding Impacts of Sea Level Rise and Tides from Storm Induced Coastal Flooding in the City of Shanghai, China," *Environmental Earth Sciences* 63 (2011): 407–14, doi.org/10.1007/s12665 -010-0787-9.

16   Zhijun Ma et al., "Rethinking China's New Great Wall," *Science* 346, no. 6212 (2014): 912–14, pubmed.ncbi.nlm.nih.gov/25414287.

17   "Fresh Water for Shanghai."

18   Jing Xuan Teng, "What's Behind China's Record Floods?," Phys.org, August 20, 2020, phys.org/news/2020-08-china.html.

Chapter 6: Dhaka and the Sundarbans

1   Steven C. Clemens et al., "Remote and Local Drivers of Pleistocene South Asian Summer Monsoon Precipitation: A Test for Future Predictions," *Science Advances* 7, no. 23 (2021), advances.sciencemag.org/content/advances/7/23 /eabg3848.full.pdf.

2   "The Sundarbans," UNESCO World Heritage Convention, accessed July 31, 2022, whc.unesco.org/en/list/798.

3   Needs Assessment Working Group, "Monsoon Floods 2020: Coordinated Preliminary Impact and Needs Assessment, Bangladesh," August 3, 2020, humanitarianresponse.info/sites/www.humanitarianresponse.info/files /documents/files/nawg_monsoon_flood_preliminary_impact_and_kin _20200802_final.pdf.

4   Md Hedait Hossain Molla, "Sundarbans Once Again Protects Bangladesh," *Dhaka Tribune*, May 21, 2020, dhakatribune.com/bangladesh/nation /2020/05/21/sundarbans-once-again-protects-the-coast.

5   Richard M. Eaton, *The Rise of Islam and the Bengal Frontier, 1204–1760* (Berkeley: University of California Press, 1993), 309, ark.cdlib.org/ark: /13030/ft067n99v9/.

6   Eaton, 20.

7   Eaton, 306. He goes on: "Similar ideas are found in Saiyid Sultan's treatment of Abraham, the supreme patriarch of Judeo-Christian-Islamic civilization. Born and raised in a forest, Abraham travelled to Palestine, where he attracted tribes from nearby lands, mobilized local labor to cut down the forest, and built a holy place, Jerusalem's Temple, where prayers could be offered to Niranjan. It is obvious that the main themes of Abraham's life as recorded by Saiyid Sultan — his sylvan origins, his recruitment of nearby tribesmen, his leadership in clearing the forest, and his building a house of prayer — precisely mirrored the careers of the hundreds of pioneering" farmers in the eastern Bengal delta.

8   Eaton, 308, 312.

9   John Taylor, *The Cotton Manufacture of Dacca: Descriptive and Historical Account of Cotton Manufacture of Dacca, in Bengal* (London: John Mortimer, 1851), 12, archive.org/stream/1851cottonmanufactureofDacca/EX .1851.212_djvu.txt.

10 Ahsanul Kabir and Bruno Parolin, "Planning and Development of Dhaka — A Story of 400 Years" (paper presented at the 15th International Planning History Society Conference, Brazil, July 15–18, 2012), researchgate.net /publication/324746990_planning_and_development_of_dhaka-a_story_of _400_years.

11 James Taylor, *A Sketch of the Topography and Statistics of Dacca* (Calcutta: G.H. Huttmann, Military Orphan Press, 1840), 365, books.google.ca /books?id=6kcOAAAAQAAJ.

12 Taylor, *A Sketch*, 367.

13 Sophie Mogg, "*Gossypium arboreum* — A Travelling Botanist: The World's Most Used Fibre!," *Herbology Manchester* (blog), Manchester Museum Herbarium, December 21, 2019, herbologymanchester.wordpress.com/tag /gossypium-arboreum.

14 Hasan Shahriar Khan, "Rethinking Buckland Bund as a Historical Street," World Architecture Community, July 23, 2009, worldarchitecture.org/architecture -projects/enge/rethinkingbucklandbundasahistoricalstreet-project-pages .html.

15 Abdul Mannan, "Chittagong — Looking for a Better Future," *New Age*, April 12, 2012.

16 Dhrubo Alam and Fatiha Polin, "Historic Overview of Transport and Urban Planning for Dhaka" (PowerPoint slides, March 2020), researchgate.net /publication/339913764_Historic_Overview_of_Transport_Urban _Planning_for_Dhaka.

17 Tarun Sarkar, "Misfortune of a Majestic Landmark: Historic Buckland Bund Stands Years of Abuse," *Daily Star*, July 31, 2020, thedailystar.net/frontpage /news/misfortune-majestic-landmark-1938865.

18 Alam and Polin, "Historic Overview," 2.

19 Jeremy Seabrook, *The Song of the Shirt: The High Price of Cheap Garments, from Blackburn to Bangladesh* (London: Hurst, 2015), 82–83.

20 Seabrook, 269.

21 Nicholas Kristof, "The Everyday Disaster of Life in Bangladesh," *New York Times*, May 19, 1991, nytimes.com/1991/05/19/weekinreview/the-everyday-disaster-of-life-in-bangladesh.html.

22 Nicholas Kristof, "What Can Biden's Plan Do for Poverty? Look to Bangladesh," *New York Times*, March 10, 2021, nytimes.com/2021/03/10/opinion/biden-child-poverty-bangladesh.html.

23 Mehedi Hasan, "Remittance Inflow to Bangladesh Accounted for 6.6% of GDP in 2020," *Dhaka Tribune*, May 18, 2021, dhakatribune.com/business/2021/05/18/remittance-inflow-to-bangladesh-accounted-for-6-6-of-gdp-in-2020.

24 CNA, the Singapore broadcaster, has done some excellent reporting on climate change in Asia. See particularly "Bangladesh's Battle with Water: Can It Save Its Climate Refugees?," *Once Upon a River*, uploaded February 20, 2021, youtube.com/watch?v=tV42sBSq1Mw&t=2287s; "Sink or Swim? Asia's Sinking Villages Engulfed by Rising Seas," *The Longest Day*, uploaded September 19, 2020, youtube.com/watch?v=hA_bCRjqo9o&t=1342s; "Surviving Drought: The Fight to Reclaim Asia's Lost Lands," *The Longest Day*, uploaded September 19, 2020, youtube.com/watch?v=NGjxnma2Y2c&t=4s.

25 "Unlocking the Potential of the Coastal Zone in Bangladesh," Dutch Water Sector, July 23, 2020, dutchwatersector.com/news/unlocking-the-potential-of-the-coastal-zone-in-bangladesh.

26 Warren Cornwall, "As Sea Levels Rise, Bangladeshi Islanders Must Decide Between Keeping the Water Out — or Letting It In," *Science*, March 1, 2018, science.org/content/article/sea-levels-rise-bangladeshi-islanders-must-decide-between-keeping-water-out-or-letting.

27 Shahnoor Hasan, Jaap Evers, and Margreet Zwarteveen, "The Transfer of Dutch Delta Planning Expertise to Bangladesh: A Process of Policy Translation," *Environmental Science and Policy* (December 2019), doi.org/10.1016/j.envsci.2019.11.001.

28 "Bangladesh Fertility Rate 1950–2022," Macrotrends, accessed July 31, 2022, macrotrends.net/countries/BGD/bangladesh/fertility-rate.

29 "Sundarbans Bears the Brunt of Yaas While Shielding Rest of Bangladesh," *Dhaka Tribune*, May 30, 2021, dhakatribune.com/bangladesh/nation/2021/05/30/sundarbans-bears-the-brunt-of-yaas-while-shielding-rest-of-bangladesh.

30 Shannon McDonagh, "Are Bangladesh's Floating Gardens the Future of Farming?," *Euronews*, April 12, 2021, euronews.com/green/2021/04/12

/bangladesh-s-floating-vegetable-gardens-are-a-path-to-the-future-of-farming. See also Kalpana Sunder, "The Remarkable Floating Gardens of Bangladesh," September 10, 2020, BBC, bbc.com/future/article/20200910-the-remarkable-floating-gardens-of-bangladesh; Bangladesh Agricultural Research Institute, "Traditional Floating Garden Practices for Seedling Production," Food and Agriculture Organization of the United Nations, December 2020, teca.apps.fao.org/teca/en/technologies/8869.

31   Mizan R. Khan et al., "High-density Population and Displacement in Bangladesh," *Science* 372, no. 6548 (June 18, 2021): 1290–93, doi.org/10.1126/science.abi6364.

32   "Which Future Will We Choose?" (simulation of Dhaka under two climate change scenarios), Climate Central, accessed July 31, 2022, picturing.climatecentral.org/location/23.762466,90.3763871.

33   Sarder Shafiqul Alam et al., "Building Climate-Resilient, Migrant-Friendly Cities and Towns," Urban Climate Change Programme, International Centre for Climate Change and Development (July 2018), icccad.net/wp-content/uploads/2018/10/Policy-Brief-on-Climate-Migration-and-Cities.pdf.

34   "In Venice, I Heard Bangla Everywhere: Amitav Ghosh on 'Gun Island,'" *The Wire*, uploaded June 19, 2019, youtube.com/watch?v=r5RbdChKMv4.

Chapter 7: The Salish Sea, Vancouver, and Seattle

1   Sammy Estfall and Amanda Coletta, "Crushing Heat Wave in Pacific Northwest and Canada Cooked Shellfish Alive by the Millions," *Washington Post*, July 8, 2021, washingtonpost.com/world/2021/07/08/canada-sea-creatures-boiling-to-death.

2   "About the Salish Sea," SeaDoc Society, accessed July 31, 2022, seadocsociety.org/about-the-salish-sea; Bert Webber, "How the Salish Sea Got Its Name," SeaDoc Society, accessed July 31, 2022, seadocsociety.org/how-the-salish-sea-got-its-name.

3   Webber, "How the Salish Sea."

4   Warren King, "Smallpox Epidemic Still Haunts Doctors," *Seattle Times*, March 5, 2003, archive.seattletimes.com/archive/?date=20030305&slug=smallpox05m. It is always interesting to have one's memory challenged for its veracity. There is no way I can check what I remember about that day at Point Defiance Park, because everyone else is long gone. But when it comes to this memory, as well as one I have of being vaccinated against smallpox, I've been able to determine just when things happened. The 6.3 magnitude earthquake on February 14, 1946, occurred at 7:18 p.m. It's possible that I had a smallpox vaccination that day because I remember my mother noting that it already was beginning to show the characteristic inflammation just before all hell broke loose. Six weeks

or so later, my second memory kicks in: being in a firehouse for a mass smallpox vaccination and watching as a nursing sister wearing a flowing coif came down the circular staircase. I thought she was an angel. That vaccination didn't "take," as I remember, but who knows? It was so long ago, and I was so little.

5 Natural Resources Canada, *Earthquakes in Southwestern British Columbia*, 2011, seismescanada.rncan.gc.ca/pprs-pprp/pubs/GF-GI/GEOFACT _earthquakes-SW-BC_e.pdf; Seattle Office of Emergency Management, *Earthquakes*, April 23, 2014, seattle.gov/Documents/Departments /Emergency/PlansOEM/SHIVA/2014-04-23_Earthquakes(0).pdf.

6 Duncan McLaren et al., "A Post-Glacial Sea Level Hinge on the Central Pacific Coast of Canada," *Quaternary Science Reviews* 97 (August 2014): 148–69, doi.org/10.1016/j.quascirev.2014.05.023; Quentin Mackie et al., "Early Environments and Archaeology of Coastal British Columbia," in *Trekking the Shore: Changing Coastlines and the Antiquity of Coastal Settlements*, eds. Nuno F. Bicho, Jonathan A. Haws, and Loren G. Davis (New York: Springer 2011): 51–103.

7 For a fascinating account of the Clovis discoveries, see Charles C. Mann, *1491: New Revelations of the Americas Before Columbus* (New York: Vintage, 2006), 164–78. Also note that radiocarbon dating has been superseded largely by accelerator mass spectrometry dating, which can be done with far smaller samples, thus removing bias that could have given later dates because the material being dated could be contaminated by more recent deposits of organic material.

8 Bridget Alex, "Monte Verde: Our Earliest Evidence of Humans Living in South America," *Discover*, October 31, 2019, discovermagazine.com/planet-earth /monte-verde-our-earliest-evidence-of-humans-living-in-south-america.

9 Lionel E. Jackson Jr. and Michael C. Wilson, "The Ice-Free Corridor Revisited," *Geotimes*, February 2004, geotimes.org/feb04/feature_Revisited.html.

10 Matthew R. Bennett et al., "Evidence of Humans in North America During the Last Glacial Maximum," *Science* 373, no. 6562 (September 24, 2021): 1528–31, doi.org/10.1126/science.abg7586.

11 Mackie et al., "Early Environments and Archaeology," 54.

12 Todd J. Braje et al., "Fladmark + 40: What Have We Learned About a Potential Pacific Coast Peopling of the Americas?," *American Antiquity* (October 2019): 1–21, doi.org/10.1017/aaq.2019.80.

13 Robert E. Kopperl et al., "The Bear Creek Site (45KI839), a Late Pleistocene–Holocene Transition Occupation in the Puget Sound Lowland, King County, Washington," *PaleoAmerica*, 1 no. 1 (January 28, 2015): 116–20, doi.org/10.1179/2055556314Z.0000000004.

14 Story collected by Arthur Ballard, adapted by Tom Dailey, as quoted in BJ Cummings, *The River That Made Seattle: A Human And Natural History of the Duwamish* (Seattle: University of Washington Press, 2020), 16–17.

15 D. McLaren et al., "Terminal Pleistocene Epoch Human Footprints from the Pacific Coast of Canada," *PLOS ONE* 13, no. 3 (March 28, 2018), doi.org/10.1371/journal.pone.0193522.

16 Roshini Nair, "Archeological Find Affirms Heiltsuk Nation's Oral History," CBC News, March 30, 2017, cbc.ca/news/canada/british-columbia/archeological-find-affirms-heiltsuk-nation-s-oral-history-1.4046088.

17 Quentin Mackie, "Raven-Walking & Geological Transformation," *Northwest Coast Archeology* (blog), December 10, 2009, qmackie.com/2009/12/10/raven-walking-geological-transformation.

18 Condensed from Eric Lloyd, "Southern Resident Killer Whales Make First Major Appearance in Salish Sea After 109 Days," *CTV News Vancouver Island*, July 28, 2021, vancouverisland.ctvnews.ca/southern-resident-killer-whales-make-first-major-appearance-in-salish-sea-after-109-days-1.5527140.

19 Nigel Haggan et al., "12,000+ Years of Change: Linking Traditional and Modern Ecosystem Science in the Pacific Northwest" (Fisheries Centre Working Paper Series 2006-02, University of British Columbia, Vancouver, 2006), data.skeenasalmon.info/dataset/545c2276-d1c4-4817-a716-dd096dd9ec08/resource/78febfbe-f6b9-49ae-84e7-893792b8b95c/download/haggen-et-al-2006-12000-years-of-change-linking-trad-modern-science.pdf.

20 Daniel DeMay, "A Brief History of the Skagit Valley Tulip Festival," *Seattle PI*, April 26, 2018, seattlepi.com/seattlenews/article/Photos-A-brief-history-of-the-Skagit-Valley-12864382.php.

21 Noel V. Bourasaw, "The Calhoun Brothers of LaConner," *Skagit River Journal of History & Folklore*, July 22, 2004, skagitriverjournal.com/WestCounty/LaConner/Pioneers/Pre1890/Calhoun/Calhoun01-FirstwSullivan.html.

22 Albert County Museum, "The Dykes of Albert County," *Community Stories*, Digital Museums Canada, accessed July 31, 2022, communitystories.ca/v1/pm_v2.php?id=story_line&lg=English&fl=0&ex=00000550&sl=4344&pos=1; "Today, August 16th, Is the Chipoudie Monument Unveiling in Riverside-Albert," Albert County Museum, August 16, 2019, albertcountymuseum.com/news/today-august-16th-is-the-chipoudie-monument-unveiling-in-riverside-albert-visit-us-to-learn-more-about-the-acadians-in-albert-county.

23 Noel V. Bourasaw, ed., "*Illustrated History of Skagit and Snohomish Counties*: Chapter 1: Period of Settlement," January 31, 2008, *Skagit River Journal of History & Folklore*, skagitriverjournal.com/scounty/library/ih06/ih06sec2chap1-1.html.

24 A treaty signed in the mid-nineteenth century with the various Indigenous groups in the U.S. portion of the Salish Sea stated, "The right of taking fish, at all usual and accustomed grounds and stations, is further secured

to said Indians in common with all other citizens of the Territory, and of erecting temporary houses for the purpose of curing them, together with the privileges of hunting, gathering roots and berries, and pasturing their horses on open and unclaimed lands." What this meant exactly wasn't determined for 119 years, when a judge ruled that tribes subject to the treaty and the settlers could each take up to 50 percent of the harvestable fish. This in turn led to more court decisions giving the Indigenous groups a say in managing the salmon stock, including fostering salmon streams and breeding grounds. See Phil Dougherty, "Boldt Decision: United States v. State of Washington," HistoryLink, last modified August 24, 2020, historylink.org/file/21084.

25 Paige Browning, "Swinomish Tribe to Sue Army Corps over Salmon," KUOW, September 13, 2021, kuow.org/stories/swinomish-tribe-intends-to -sue-army-corps-over-salmon-loss.

26 Dick Clever, "Fish vs. Farms on the Skagit Delta," *Seattle Weekly*, August 14, 2012, seattleweekly.com/news/fish-vs-farms-on-the-skagit -delta; Western Washington Agricultural Association, NOAA's National Marine Fisheries Service, and Washington Department of Fish and Wildlife, "Skagit Delta Tidegates and Fish Initiative Implementation Agreement," May 28, 2008, salishsearestoration.org/images /archive/0/07/20200912010303%21WWAA_et_al_2008_tidegates_and _fish_initiative_agreeement.pdf.

27 City of Surrey, "Coastal Flood Adaptation Strategy," November 2019, surrey.ca/sites/default/files/media/documents /CFASFinalReportNov2019.pdf; Tracy Holmes, "$76 Million Pledged for Coastal Flooding Mitigation in Surrey and Delta," *Peace Arch New*, May 24, 2019, peacearchnews.com/news/76-million-pledged-for-coastal-flooding -mitigation-in-south-surrey-and-delta.

28 The Columbia, by way of comparison, is 1,243 miles (2,000 kilometres) long.

29 British Columbia Ministry of Environment, *Sea Level Rise Adaptation Primer: A Toolkit to Build Adaptive Capacity on Canada's South Coasts*, Fall 2013, 94, gov.bc.ca/assets/gov/environment/climate-change/adaptation /resources/slr-primer.pdf.

30 "Flood Protection," City of Richmond, June 15, 2022, richmond.ca/services /rdws/dikes.htm.

31 Kerr Wood Leidal, *Flood Protection Management Strategy Land Raising Review: Final Report*, November 30, 2020. Found as Attachment 1 in the agenda of the March 15, 2021, meeting of the Richmond General Purposes Committee, docplayer.net/204226278-General-purposes-committee -engineering-and-public-works-division.html. The report was received, but action not taken immediately.

32 Kerr Wood Leidal, 4-2–4-3.

33 "Which Future Will We Choose?" (simulation of Vancouver under two climate change scenarios), Climate Central, accessed July 31, 2022, picturing.climatecentral.org/location/49.2767851,-123.1121189; Coastal Risk Screening Tool (map for False Creek area, Vancouver), Climate Central, accessed July 31, 2022, tinyurl.com/h5fkzjbk.

34 City of Vancouver, *Vancouver's Changing Shoreline: Preparing for Sea Level Rise*, December 2018, vancouver.ca/files/cov/vancouvers-changing-shoreline.pdf.

35 "Projected Impacts of Sea Level Rise" (map), City of Seattle GIS, November 18, 2020, seattlecitygis.maps.arcgis.com/home/webmap/viewer.html?webmap =5371d5f1794647baaccf35cde47173bb.

36 This is the premise behind BJ Cummings's *The River That Made Seattle: A Human Land Natural History of the Duwamish.*

37 Cummings, *The River That Made Seattle*, 39.

38 Cummings, 70.

39 Alan J. Stein, "Howard A. Hanson Dam," HistoryLink.org, September 10, 2009, historylink.org/file/3549.

40 "Lower Duwamish Waterway, Seattle, WA: Cleanup Activities," U.S. Environmental Protection Agency, accessed July 31, 2022, cumulis.epa .gov/supercpad/SiteProfiles/index.cfm?fuseaction=second.cleanup&id =1002020.

41 Jeffrey Lin, "Visualizing Sea Level Rise on the Duwamish River," *GIS & You* (blog), March 4, 2020, gisandyou.org/2020/03/04/sea-level-rise -duwamish-river.

42 Ashli Blow, "Climate Change Takes a Toll on Seattleites' Mental Health," Crosscut, June 6, 2022, crosscut.com/environment/2022/06/climate -change-takes-toll-seattleites-mental-health.

43 U.S. Army Corps of Engineers, Seattle District, *Alki Coastal Erosion Control Project: Integrated Detailed Project Report and Final Environmental Assessment*, September 2019, nws.usace.army.mil/Portals/27/docs/environmental/resources /2018EnvironmentalDocuments/FINAL%20Integrated%20DPR-EA%20Alki %20Sect%20103_12Sep2019.pdf?ver=kC29L5sUAHXH3eUSO0ZeHA %3d%3d.

Chapter 8: What Can Be Done — What Will Be Done?

1 View the video at youtube.com/watch?v=9PrbLsQ_g7s.

2 Dio Suhenda, "Govt Says Capital City Relocation Will Resume," *Jakarta Post*, August 28, 2021, thejakartapost.com/paper/2021/08/27/govt-says -capital-city-relocation-will-resume.html; Vincent Fabian Thomas, "New Capital Will Take Decades, Not Years: Bappenas," *Jakarta Post*, September

4, 2021, thejakartapost.com/business/2021/09/03/new-capital-will-take
-decades-not-years-bappenas.html.

3   "Carbon Dioxide Now More Than 50% Higher Than Pre-Industrial Levels,"
National Oceanic and Atmospheric Association, June 3, 2022, noaa.gov/news
-release/carbon-dioxide-now-more-than-50-higher-than-pre-industrial-levels.

4   New Climate Institute and Climate Analytics, *Global Reaction to Energy
Crisis Risks Zero Carbon Transition: Analysis of Government Responses
to Russia's Invasion of Ukraine*, June 2022, Climate Action Tracker,
climateactiontracker.org/documents/1055/CAT_2022-06-08_Briefing
_EnergyCrisisReaction.pdf.

5   Intergovernmental Panel on Climate Change, *Climate Change 2022:
Mitigation of Climate Change*, accessed July 31, 2022, ipcc.ch/report/ar6/wg3.
For a readable assessment see Jeff Godell, "There's a Silver Lining to the U.N.'s
Final Warning of Climate Change," *Rolling Stone*, April 4, 2022, rollingstone
.com/politics/politics-features/ipcc-report-climate-change-warning-1335131.

6   IPCC, "The Evidence Is Clear: The Time for Action Is Now. We Can
Halve Emissions by 2030," news release, April 4, 2022, ipcc.ch/2022/04/04
/ipcc-ar6-wgiii-pressrelease.

7   See Mann, *1491*, particularly pages 352–66, for a gripping account of the
population crash.

8   Robert A. Dull et al., "The Columbian Encounter and the Little Ice Age:
Abrupt Land Use Change, Fire, and Greenhouse Forcing," *Annals of the
Association of American Geographers* 100, no. 4 (2010): 755–71, doi.org/10
.1080/00045608.2010.502432.

9   Mike Gaworecki, "Is Planting Trees as Good for the Earth as Everyone
Says?," *Mongoba*, May 13, 2021, news.mongabay.com/2021/05/is-planting
-trees-as-good-for-the-earth-as-everyone-says; "Reforestation and Forest
Restoration," Conservation Effectiveness, conservationeffectiveness.org
/datavis/?strategy=13.

10 H.D. Le et al., "More Than Just Trees: Assessing Reforestation Success in
Tropical Developing Countries," *Journal of Rural Studies* 28, no. 1 (2011):
5–19, doi.org/10.1016/j.jrurstud.2011.07.006.

11 George Monbiot, "Indonesia Is Burning. So Why Is the World Looking
Away?," *Guardian*, October 30, 2015, theguardian.com/commentisfree/2015
/oct/30/indonesia-fires-disaster-21st-century-world-media.

12 Chris Mooney et al., "Countries' Climate Pledges Built on Flawed
Data, *Post* Investigation Finds," *Washington Post*, November 7, 2021,
washingtonpost.com/climate-environment/interactive/2021/greenhouse
-gas-emissions-pledges-data/?itid=hp-top-table-main.

13 Christian Hoffmann, Michel Van Hoey, and Benedikt Zeumer,
"Decarbonization Challenge for Steel," McKinsey and Company, June 3, 2020,

mckinsey.com/industries/metals-and-mining/our-insights/decarbonization
-challenge-for-steel.

14 Austin Himes and Gwen Busby, "Wood Buildings as a Climate Solution,"
Developments in the Built Environment 4 (November 2020), doi.org/10.1016/j
.dibe.2020.100030.

15 For why street trees are important, see Carly Cassella, "Even a Single Tree
Can Help Cool Down a City at Nighttime. Here's How," *Science Alert*,
July 18, 2021, sciencealert.com/a-single-tree-can-help-cool-down-a-city
-in-the-evening-study-finds.

16 Agence France-Presse, "World's Biggest Machine Capturing Carbon from
Air Turned On in Iceland," *Guardian*, September 9, 2021, theguardian.com
/environment/2021/sep/09/worlds-biggest-plant-to-turn-carbon-dioxide-into
-rock-opens-in-iceland-orca.

17 "The Thames Barrier," Gov.UK, May 16, 2022, gov.uk/guidance/the-thames
-barrier.

18 Jane Sidell, "Archaeology and the London Thames: Past, Present and Future,"
Archeology International, October 23, 2001, doi.org/10.5334/ai.0505.

19 Diccon Hart et al., "Early Neolithic Trackways in the Thames Floodplain
at Belmarsh, London Borough of Greenwich," *Proceedings of the Prehistoric
Society* 81 (December 2015): 1–23.

20 F.C.J. Spurrell, "Early Sites and Embankments on the Margins of the Thames
Estuary," *Archaeological Journal* 42, no. 1 (1885): 269–302, doi.org/10.1080
/00665983.1885.10852177.

21 "Explosion of Gunpowder Magazines at Erith, 1864," *Where Thames Smooth
Waters Glide* (blog), accessed July 31, 2022, thames.me.uk/explosion1864.htm.

22 Antonio Brambati et al., "The Lagoon of Venice: Geological Setting, Evolution
and Land Subsidence," *Episodes* 26, no. 3 (2003): 264–68, doi.org/10
.18814/epiiugs/2003/v26i3/020.

23 There are several fascinating histories of Venice for the general reader. I
found *Voyages dans l'histoire venise* by Peter Mentzel, translated by Gari
Ulubeyan (Paris: NG France, 2011), particularly interesting and informative.

24 Marcella Hazan, "Marcella Hazan's Bolognese Sauce," *New York Times*,
accessed July 29, 2022, cooking.nytimes.com/recipes/1015181-marcella
-hazans-bolognese-sauce.

25 Anna Momigliano, "Venice Tourism May Never Be the Same. It Could Be
Better," *New York Times*, July 2, 2020, nytimes.com/2020/07/02/travel/venice
-coronavirus-tourism.html. Note, in 2022 a tourist entry fee will be charged,
varying from three to ten euros depending on the season. Large cruise ships
are now also banned from sailing into the lagoon; Chloe Taylor, "Venice Set
to Introduce an Entry Fee and Booking System for Tourists," *CNBC*, August

23, 2021, cnbc.com/2021/08/23/venice-set-to-introduce-an-entry-fee-and
-booking-system-for-tourists.html.

26 Amitav Ghosh, *The Great Derangement: Climate Change and the Unthinkable* (Chicago: University of Chicago Press, 2017).

27 Amitav Ghosh, *Gun Island* (Toronto: Penguin Random House Canada, 2019), 218.

28 Olivia Rosane, "Venice Floods After New Barrier Fails to Activate," *Guardian*, December 9, 2020, theguardian.com/weather/2020/dec/08/venice-floods-as-forecasts-fail-to-predict-extent-of-high-tide.

29 "Can Dutch-Promoted 'Sponge Cities' Be the Answer to Frequent Flooding in Urban China?," *Shanghai Daily*, October 17, 2016, archive.shine.cn/opinion/foreign-perspectives/Can-Dutchpromoted-sponge-cities-be-the-answer-to-frequent-flooding-in-urban-China/shdaily.shtml.

30 Jeroen Rijke et al., "Room for the River: Delivering Integrated River Basin Management in the Netherlands," *International Journal of River Basin Management* 10 (2012): 369–82, doi.org/10.1080/15715124.2012.739173.

31 "Rijkswaterstaat: Maatregelen na 1995 voorkwamen nu erger Limburgs leed [Rijkswaterstaat: Measures After 1995 Now Prevented Worse Limburg Suffering]," *NU.nl*, trans. Google Translate, July 22, 2021, nu.nl/wateroverlast-limburg/6146916/rijkswaterstaat-maatregelen-na-1995-voorkwamen-nu-erger-limburgs-leed.amp.

32 "Europe Flooding Deaths Pass 125, and Scientists See Fingerprints of Climate Change," *New York Times*, July 16, 2021, nytimes.com/live/2021/07/16/world/europe-flooding-germany.

33 For technical details and fetching photos, see Elizabeth L. Newhouse, ed., *The Builders: Marvels of Engineering* (New York: National Geographic Society, 1992): 196–99.

34 Michael Kimmelmann, "The Dutch Have Solutions to Rising Seas. The World Is Watching," *New York Times*, June 15, 2017, nytimes.com/interactive/2017/06/15/world/europe/climate-change-rotterdam.html.

35 Angela Moore, "To Protect a Community from Climate Change, New York Is Elevating a Park," *Reuters*, May 21, 2021, reuters.com/business/environment/protect-community-climate-change-new-york-is-elevating-park-2021-05-21; Benjamin H. Strauss et al., "Economic Damages from Hurricane Sandy Attributable to Sea Level Rise Caused by Anthropogenic Climate Change," *Nature Communications* 12, no. 2720 (May 18, 2021), doi.org/10.1038/s41467-021-22838-1.

36 Christopher Flavelle et al., "Overlapping Disasters Expose Harsh Climate Reality: The U.S. Is Not Ready," *New York Times*, September 2, 2021, nytimes.com/2021/09/02/climate/new-york-rain-floods-climate-change.html.

37 *Dutch Proposal to Dam the North Sea*, produced by Shirvan Neftchi, uploaded February 24, 2020, on CaspianReport, youtube.com/watch?v=neFMunVEE8E.

38 Sjoerd Groeskamp and Joakim Kjellsson, "NEED: The Northern European Enclosure Dam for If Climate Change Mitigation Fails," *Bulletin of the American Meteorological Society* 101, no. 7 (July 2020): E1174–89, journals .ametsoc.org/view/journals/bams/101/7/bamsD190145.xml.

39 Sjoerd Groeskamp, interviewed by author via Zoom, September 27, 2021.

40 Tom Reynolds, "Final Report of Hanford Thyroid Disease Study Released," *Journal of the National Cancer Institute* 94, no. 14 (July 17, 2002): 1046–48, doi.org/10.1093/jnci/94.14.1046.

41 Pushker A. Kharecha and James E. Hansen, "Prevented Mortality and Greenhouse Gas Emissions from Historical and Projected Nuclear Power," *Environmental Science and Technology* 47, no. 9 (March 2013): 4889–95, dx.doi.org/10.1021/es3051197.

42 Joshua S. Goldstein and Staffan A. Qvist, *A Bright Future: How Some Countries Have Solved Climate Change and the Rest Can Follow* (New York: Public Affairs, 2019).

43 Kharecha and Hansen, 4889.

44 David Stanway, Philip Wen, and Stella Qiu, "A Pollution Crackdown Compounds Slowdown Woes in China's Heartland," *Reuters*, May 24, 2019, reuters.com/article/us-china-economy-henan-pollution-insight -idUSKCN1SU025.

45 Richard Gray, "The True Toll of the Chernobyl Disaster," *BBC News*, July 25, 2019, bbc.com/future/article/20190725-will-we-ever-know -chernobyls-true-death-toll.

46 Roger J. Levin et al., "Incidence of Thyroid Cancer Surrounding Three Mile Island Nuclear Facility: The 30-Year Follow-Up," *Laryngoscope* 123, no. 8 (January 31, 2013): 2064–71, doi.org/10.1002/lary.23953.

47 *The Fukushima Daiichi Accident: Report by the Director General* (Vienna: International Atomic Energy Agency, 2015), www-pub.iaea.org/mtcd /publications/pdf/pub1710-reportbythedg-web.pdf. See also "Japan Confirms First Fukushima Worker Death from Radiation," *BBC News*, September 5, 2018, bbc.com/news/world-asia-45423575.

48 Laurence Peter, "Ukraine War: Chernobyl Scarred by Russian Troops' Damage and Looting," *BBC News*, June 3, 2022, bbc.com/news /world-europe-61685643 and Hugo Bachega, "Ukraine War: Bombardments Near Nuclear Plant a Concern for All — UN Chief," *BBC News*, August 19, 2022, bbc.com/news/world-europe-62608529.

49 Hannah Ritchie, "What Are the Safest and Cleanest Sources of Energy?," Our World in Data, February 10, 2020, ourworldindata.org/safest-sources

-of-energy. Note that Goldstein and Qvist use deaths/terawatt-hour in their book also.

50 Kerstine Appunn, "The History Behind Germany's Nuclear Phase-Out," *Clean Energy Wire*, March 9, 2021, cleanenergywire.org/factsheets/history -behind-germanys-nuclear-phase-out.

51 Mycle Schneider, *The World Nuclear Industry Status Report 2021* (Paris: Mycle Schneider, 2021), worldnuclearreport.org/IMG/pdf/wnisr2021-lr .pdf.

52 Stephen Jarvis, Olivier Deschenes, and Akshaya Jha, "The Private and External Costs of Germany's Nuclear Phase-Out" (Working Paper 26598, National Bureau of Economic Research, December 2019), nber.org/papers/w26598.

53 Sabine Kinkartz, "Ukraine Crisis Forces Germany to Change Course on Energy," *DW*, March 1, 2022, dw.com/en/ukraine-crisis-forces-germany-to -change-course-on-energy/a-60968585.

54 James Conca, "China Will Lead the World in Nuclear Energy, Along With All Other Energy Sources, Sooner Than You Think," *Forbes*, April 23, 2021, forbes.com/sites/jamesconca/2021/04/23/china-will-lead-the-world-in -nuclear-energy-along-with-all-other-energy-sources-sooner-than-you-think /?sh=69725f72778c.

55 "Nuclear Explained," U.S. Energy Information Administration, last modified April 18, 2022, eia.gov/energyexplained/nuclear/us-nuclear-industry.php and "Nuclear Power Plants," Canadian Nuclear Safety Commission, last modified March 2, 2022, cnsc-ccsn.gc.ca/eng/reactors/power-plants/index.cfm.

56 Goldstein and Qvist, 88.

57 Izzy Kapnick, "Feds Hit Turkey Point Nuclear Plant With Performance Downgrade," *Miami New Times*, May 17, 2021, miaminewtimes.com /news/nrc-downgrades-turkey-point-nuclear-plant-12255500; "Turkey Point Licensed for 80 Years of Operation," *World Nuclear News*, December 6, 2019, world -nuclear-news.org/Articles/Turkey-Point-licensed-for-80-years-of-operation.

58 Calculated from data at "Florida," U.S. Energy Information Administration, accessed July 31, 2022, eia.gov/state/?sid=FL.

59 John Vidal, "Are Coastal Nuclear Power Plants Ready for Sea Level Rise?," *Hakai Magazine*, August 21, 2018, hakaimagazine.com/features/are-coastal -nuclear-power-plants-ready-for-sea-level-rise.

60 Rebecca Tuhus-Dubrow, "The Activists Who Embrace Nuclear Power," *New Yorker*, February 19, 2021, newyorker.com/tech/annals-of-technology /the-activists-who-embrace-nuclear-power.

61 See Ivan Semeniuk, "Magnet Breakthrough Brings Fusion Energy Closer to Reality," *Globe and Mail*, September 8, 2021, theglobeandmail.com/business /article-magnet-breakthrough-brings-fusion-energy-closer-to-reality/; Ivan Semeniuk, "Laser Fusion Experiment Nears Crucial Break Even Point in Energy

Generation," *Globe and Mail*, August 19, 2021, theglobeandmail.com/canada/article-laser-fusion-experiment-nears-crucial-break-even-point-in-energy.

62 Jon Huang, Claire O'Neill, and Hiroko Tabuchi, "Bitcoin Uses More Electricity Than Many Countries. How Is That Possible?," *New York Times*, September 3, 2021, nytimes.com/interactive/2021/09/03/climate/bitcoin-carbon-footprint-electricity.html.

63 "Bitcoin (BTC / CAD)," Google Finance, google.com/finance/quote/BTC-CAD.

64 James T. Areddy, "China Creates Its Own Digital Currency, a First for Major Economy," *Wall Street Journal*, April 5, 2021, wsj.com/articles/china-creates-its-own-digital-currency-a-first-for-major-economy-11617634118.

65 "China Declares All Crypto-Currency Transactions Illegal," *BBC News*, September 24, 2021, bbc.com/news/technology-58678907. The BBC reports that "trading crypto-currency has officially been banned in China since 2019, but continued online through foreign exchanges." In June 2021, the government "told banks and payment platforms to stop facilitating transactions and issued bans on 'mining' the currencies." See also Peter Hoskins, "China Power Cuts: What Is Causing the Country's Blackouts?," *BBC News*, September 30, 2021, bbc.com/news/business-58733193.

66 Andrew Ross Sorkin et al., "Miami Wants to Be the Hub for Bitcoin," *New York Times*, March 23, 2021, nytimes.com/2021/03/23/business/dealbook/miami-suarez-crypto.html; Mackenzie Sigolas, "El Salvador Has Just Started Mining Bitcoin Using the Energy from Volcanoes," *CNBC*, October 1, 2021, cnbc.com/2021/10/01/el-salvador-just-started-mining-bitcoin-with-volcanoes-for-the-first-time-ever-and-theyve-already-made-269.html.

67 Glen McGregor, "Poilievre Personally Holds Investment in Bitcoin as He Promotes Crypto to Canadians," *CTV News*, May 17, 2022, ctvnews.ca/politics/poilievre-personally-holds-investment-in-bitcoin-as-he-promotes-crypto-to-canadians-1.5907615.

68 Delton B. Chen, Joel van der Beek, and Jonathan Cloud, "Climate Mitigation Policy as a System Solution: Addressing the Risk Cost of Carbon," *Journal of Sustainable Finance & Investment* 7, no. 3 (2017): 233–74, doi.org/10.1080/20430795.2017.1314814.

69 Bill McKibben, "It's Not Science Fiction," *New York Review*, December 17, 2020, nybooks.com/articles/2020/12/17/kim-stanley-robinson-not-science-fiction.

70 Ryan Browne, "Bill Gates Says Crypto and NFTS Are '100% Based on Greater Fool Theory,'" *CNBC*, June 15, 2022, cnbc.com/2022/06/15/bill-gates-says-crypto-and-nfts-are-based-on-greater-fool-theory.html.

71 Billy Bambrough, "Bill Gates Issues Serious Bitcoin Warning as Tesla Billionaire Elon Musk Stokes Crypto Price 'Mania,'" *Forbes*, February

26, 2021, forbes.com/sites/billybambrough/2021/02/26/bill-gates-issues
-serious-bitcoin-warning-as-tesla-billionaire-elon-musk-stokes-crypto-price
-mania/?sh=2c7aceb23892.

72 Gates, *How to Avoid*, 176–77.

73 David Keith, "What's the Least Bad Way to Cool the Planet?," *New York Times*, October 1, 2021, nytimes.com/2021/10/01/opinion/climate-change -geoengineering.html.

74 Elizabeth Kolbert, *Under a White Sky: The Nature of the Future* (New York: Penguin Random House, 2021), 59.

75 Patricia Mazzei, "A 20-Foot Sea Wall? Miami Faces the Hard Choices of Climate Change," June 2, 2021, nytimes.com/2021/06/02/us/miami-fl-seawall -hurricanes.html; see also Anne Barnard, "The $119 Billion Sea Wall That Could Defend New York ... or Not," *New York Times*, January 17, 2020, nytimes .com/2020/01/17/nyregion/the-119-billion-sea-wall-that-could-defend-new -york-or-not.html.

76 See my *Concrete: From Ancient Origins to a Problematic Future* (Regina: University of Regina Press, 2020).

77 Tifa Asrianti, "Protecting the Country Through Restoration: Jokowi Plants Mangroves with Local Community," *Jakarta Post*, October 1, 2021, thejakartapost.com/news/2021/10/01/protecting-the-country-through -restoration-jokowi-plants-mangroves-with-local-community.html.

78 *Green City: People, Nature and Urban Place* (Montreal: Véhicule Press, 2008), in which Kochi is one of ten cities whose relationship with the environment is examined.

79 Dr. U.K. Gopalan, Kochi, Kerala State, India, interviewed by author February 21, 2005.

80 Antonio B. Rodriguez et al., "Oyster Reefs Can Outpace Sea-Level Rise," *Nature Climate Change* 4 (2014): 493–97, doi.org/10.1038/nclimate2216.

81 Eric Klinenberg, "The Seas Are Rising. Could Oysters Help?," *New Yorker*, August 2, 2021, newyorker.com/magazine/2021/08/09/the-seas-are-rising-could -oysters-protect-us.

82 "What is a living shoreline?," *National Ocean Service*, oceanservice.noaa.gov /facts/living-shoreline.html.

83 Chad Shmukler, "Stilt Houses of Texas," *Hatch*, December 5, 2018, hatchmag.com/articles/stilt-houses-texas/7714730.

84 S. Biswas, M.A. Hasan, and M. S. Islam, "Stilt Housing Technology for Flood Disaster Reduction in the Rural Areas of Bangladesh," *International Journal of Research in Civil Engineering, Architecture & Design* 3, no. 1 (January–March 2015): 1–6, researchgate.net/publication/276883175 _Stilt_Housing_Technology_for_Flood_Disaster_Reduction_in_the_Rural _Areas_of_Bangladesh.

85   Nambi Appadurai, "South Asia Confronts a Double Disaster: Cyclone and COVID-19," World Resources Institute, May 29, 2020, wri.org/insights /south-asia-confronts-double-disaster-cyclone-and-covid-19; IFRC, "Bangladesh: Cyclone YAAS — Operation Update Report, DREF Operation n°: MDRBD027 update n° 1," ReliefWeb, June 25, 2021, reliefweb.int/report/bangladesh/bangladesh-cyclone-yaas-operation-update -report-dref-operation-n-mdrbd027-update-n; "Sundarbans Bear the Brunt of Yaas While Shielding the Rest of Bangladesh," *Dhaka Tribune*, May 30, 2021, dhakatribune.com/bangladesh/nation/2021/05/30/sundarbans -bears-the-brunt-of-yaas-while-shielding-rest-of-bangladesh.

86   Ghosh, *The Hungry Tide*, 320.

87   Mike Clary, "Stilt Homes in Florida Keys Stand Tall in Face of Hurricane Irma's Fury," *Sun Sentinel*, September 15, 2017, sun-sentinel.com/news /weather/hurricane/fl-reg-hurricane-irma-stilt-homes-20170914-story.html.

88   Amanda Kolson Hurley, "The House of the Future Is Elevated," *City Lab* (blog), Bloomberg, December 8, 2017, bloomberg.com/news/articles /2017-12-08/the-high-cost-of-flood-proofing-homes.

89   Christopher Flavelle, "The Cost of Insuring Expensive Waterfront Homes Is About to Skyrocket," *New York Times*, September 24, 2021, nytimes.com/2021/09/24/climate/federal-flood-insurance-cost.html; Adam Malik, "How Home Insurance Rates in Canada Are Trending," *Canadian Underwriter*, June 8, 2021, canadianunderwriter.ca/insurance /how-home-insurance-rates-in-canada-are-trending-1004208877.

90   Gorky Bakshi, "Arcadia Education Project in Bangladesh Wins Aga Khan Architecture Award," Jagran Josh, November 26, 2019, jagranjosh.com/current -affairs/arcadia-education-project-in-bangladesh-wins-aga-khan-architecture -award-1574751067-1. Another floating school project, this one in Lagos, Nigeria, has not been as successful. Opened in 2012 to much favourable publicity, it proved to be unsafe and was eventually destroyed in 2016 by a storm; Cynthia Okoroafor, "Does Makoko Floating School's Collapse Threaten the Whole Slum's Future?," *Guardian*, June 10, 2016, theguardian.com/cities/2016/jun/10/makoko-floating-school-collapse -lagos-nigeria-slum-water.

91   Freya Sawbridge, "Are Floating Homes the Way of the Future in the Netherlands?," *DutchReview*, November 5, 2021, dutchreview.com/culture/are -floating-homes-the-way-of-the-future.

92   Fang Block, "A 'Floating Mansion' in Miami Is for Sale for $5.5 Million," *Mansion Global*, July 13, 2020, mansionglobal.com/articles /a-floating-mansion-in-miami-is-for-sale-for-5-5-million-217504.

93   "Blue Development," Arkup, accessed August 21, 2022, arkup.com/arkup40/.

94 "St. Lawrence River," *Britannica*, last updated April 1, 2022, britannica.com/place/Saint-Lawrence-River.

95 Rémy Bourdillon, "Sainte-Flavie ou le dilemma du littoral," *Un Point Cinq*, December 3, 2018, unpointcinq.ca/habitat/erosion-cotiere-sainte-flavie.

96 "Programme d'accèss à la propriété – suite," Municipalité de Sainte-Flavie, February 10, 2022, sainte-flavie.net/28/programme-d-acces-a-la-propriete/nouvelle.html.

97 "11 Facts About Hurricane Sandy," DoSomething.org, accessed July 31, 2022, dosomething.org/us/facts/11-facts-about-hurricane-sandy.

98 Ula Ilnytzky, "Four Years After Sandy, Some Places at Shore Changed Forever," *Morning Call*, October 28, 2016, mcall.com/news/nation-world/mc-nj-superstorm-sandy-four-years-later-20161028-story.html.

99 Samantha Maldonado, "City Eyes New Push to Buy Out Flood-Prone Houses as Climate Change Hits Home," *The City*, October 26, 2021, thecity.nyc/2021/10/26/22747880/nyc-buy-out-flood-prone-homes-climate-change-sandy-ida.

100 Martin Wisckol, "State Sea-Level Rise Laws Advance as Urgency Surges," *Los Angeles Daily News*, September 25, 2021, dailynews.com/2021/09/24/state-sea-level-rise-laws-advance-as-urgency-surges; California Legislature, Senate, *Sea Level Rise Revolving Loan Program* S83, 2021–2022 Regular Session, introduced in Senate December 15, 2020, leginfo.legislature.ca.gov/faces/billTextClient.xhtml?bill_id=202120220SB83.

101 Katharine J. Mach and A.R. Siders, "Reframing Strategic, Managed Retreat for Transformative Climate Adaptation," *Science* 372, no. 6548 (2021): 1294–99, doi.org/10.1126/science.abh1894.

102 Karen B. Roberts, "Managed Retreat: A Must in the War Against Climate Change," *Science Daily*, June 18, 2021, sciencedaily.com/releases/2021/06/210618091642.htm.

103 Mach and Siders, "Reframing Strategic, Managed Retreat."

104 National Round Table on the Environment and the Economy, *Paying the Price: The Economic Impacts of Climate Change for Canada* (Climate Prosperity Series Report 04, Ottawa, 2011), nrt-trn.ca/wp-content/uploads/2011/09/paying-the-price.pdf.

105 Zakir Hossain Chowdhury, "In Pictures: Bangladesh's 'Hanging Village' Is About to Drown," *TRT World Magazine*, July 8 2021, trtworld.com/magazine/in-pictures-bangladesh-s-hanging-village-is-about-to-drown-48208.

106 Parag Khanna, *Move: The Forces Uprooting Us* (Toronto: Scribner, 2021).

107 Khanna, 48.

108 This is a question I deal with more fully in the chapter on the differences and similarities between the United States and Canada in *Frenemy Nations: Love*

and Hate Between Neighbo(u)ring States (Regina: University of Regina Press, 2017).

109 Karen Gilchrist, "'There's No Higher Ground for Us': Maldives' Environment Minister Says Country Risks Disappearing," *CNBC*, May 18, 2021, cnbc.com /2021/05/19/maldives-calls-for-urgent-action-to-end-climate-change-sea -level-rise.html.

110 Kendra Pierre-Louis, "Want to Escape Global Warming? These Cities Promise Cool Relief," *New York Times*, April 15, 2019, nytimes.com/2019/04/15 /climate/climate-migration-duluth.html.

111 See Lawrence Hill's *The Book of Negroes* (New York: HarperCollins, 2007), which in some countries is published as *Somebody Knows My Name*, for an amazing story of endurance and cultural transmission.

112 And to make sure the story doesn't die as this world is turned upside down by climate change, Ghosh recently published a retelling of part of the work in verse in English: *Jungle Nama: A Story of the Sundarban* (New York: Harper Collins, 2020).

113 Omar El Akkad, *American War* (New York: Knopf, 2017), 434.

Chapter 9: The Rainbow

1 Find the video at youtube.com/watch?v=03GpPfOsFkQ.

2 David Marchese, "Yo-Yo Ma and the Meaning of Life," *New York Times*, November 20, 2020, nytimes.com/interactive/2020/11/23/magazine/yo-yo -ma-interview.html.

3 A. Muh. Ibnu Aqil and Rifki Nurfajri, "Jakarta Plans to Ban Some Groundwater Extraction. Not So Fast, Experts Say," *Jakarta Post*, October 13, 2021, thejakartapost.com/news/2021/10/13/jakarta-plans-to-ban-some -groundwater-extraction-not-so-fast-experts-say.html.

4 Kornelius Purba, "When Governor Anies Becomes a Nobody Next Year," *Jakarta Post*, February 22, 2021, thejakartapost.com/academia/2021/02/21 /commentary-when-governor-anies-becomes-a-nobody-next-year.html.

5 Brad Plumer and Nadja Popovich, "Yes, There Has Been Progress on Climate. No, It's Not Nearly Enough," *New York Times*, October 25, 2021, nytimes. com/interactive/2021/10/25/climate/world-climate-pledges-cop26.html.

6 Editorial Board, "Coal and Climate Justice," *Jakarta Post*, November 15, 2021, thejakartapost.com/paper/2021/11/14/coal-and-climate-justice.html.

7 Hans Nicholas Jong, "COP26 Cop-Out? Indonesia's Clean Energy Pledge Keeps Coal Front and Center," *Mongabay*, November 10, 2021, news .mongabay.com/2021/11/cop26-cop-out-indonesias-clean-energy-pledge -keeps-coal-front-and-center; David Vetter, "'The End of Coal': COP26 Forges New Global Agreement to Retire Dirtiest Fossil Fuel," *Forbes*,

November 3, 2021, forbes.com/sites/davidrvetter/2021/11/03/the
-end-of-coal-cop26-forges-new-global-agreement-to-retire-dirtiest-fossil-fuel
/?sh=3988e0bb1c48; and Roger Harrabin, "COP26 Climate Change Summit:
So Far, So Good-ish," BBC News, November 3, 2021, bbc.com/news/science
-environment-59150807.

8   Frank Jordons and Jill Lawless, "Leaders Vow to Protect Forests,
Plug Methane Leaks at COP 26," AP News, November 2, 2021,
apnews.com/article/climate-science-business-united-nations-scotland
-a96c50c03653ea95139f4cef3b621c70.

9   "China Says EU's Carbon Border Tax Violates Trade Principles," *Reuters*,
July 26, 2021, reuters.com/business/sustainable-business/china-says-ecs
-carbon-border-tax-is-expanding-climate-issues-trade-2021-07-26.

10   Vincent Fabian Thomas, "Fossil Fuel Subsidies Will Likely Render Carbon
Tax Useless: CPI," *Jakarta Post*, October 29, 2021, thejakartapost.com
/news/2021/10/29/fossil-fuel-subsidies-will-likely-render-carbon-tax-useless
-cpi.html.

11   Tim Figures et al., "The EU's Carbon Border Tax Will Redefine Global
Value Chains," Boston Consulting Group, October 12, 2021, bcg.com/en
-ca/publications/2021/eu-carbon-border-tax.

12   George Monbiot, "After the Failure of COP26, There's Only One Hope
for Our Survival," *Guardian*, November 14, 2021, theguardian.com
/commentisfree/2021/nov/14/cop26-last-hope-survival-climate-civil
-disobedience.

13   Simon Sharpe and Timothy M. Lenton, "Upward-Scaling Tipping Cascades
to Meet Climate Goals: Plausible Grounds for Hope," *Climate Policy* 21, no.
4 (2021): 421–33, doi.org/10.1080/14693062.2020.1870097.

14   Isabelle Gerretsen, "India's Vulnerability to Coal Shocks Exposed Amid
Surging Energy Demand and Prices," *Climate Home News*, May 10, 2021,
climatechangenews.com/2021/10/05/indias-vulnerability-coal-shocks
-exposed-amid-surging-energy-demand-prices.

SELECTED BIBLIOGRAPHY

Reports

British Columbia Ministry of Environment. *Sea Level Rise Adaptation Primer: A Toolkit to Build Adaptive Capacity on Canada's South Coasts.* Fall 2013. gov.bc.ca /assets/gov/environment/climate-change/adaptation/resources/slr-primer.pdf.

Intergovernmental Panel on Climate Change. *Climate Change 2022: Mitigation of Climate Change — Summary for Policymakers.* April 4, 2022. reliefweb.int/report /world/climate-change-2022-mitigation-climate-change-summary -policymakers.

National Round Table on the Environment and the Economy. *Paying the Price: The Economic Impacts of Climate Change for Canada.* Climate Prosperity Series Report 04, Ottawa, 2011. nrt-trn.ca/wp-content/uploads/2011/09/paying -the-price.pdf.

Needs Assessment Working Group. "Monsoon Floods 2020: Coordinated Preliminary Impact and Needs Assessment, Bangladesh." August 3, 2020. humanitarianresponse.info/sites/www.humanitarianresponse.info/files/ documents/files/nawg_monsoon_flood_preliminary_impact_and_kin _20200802_final.pdf.

Books and Articles

Alam, Sarder Shafiqul, Saleemul Huq, Faisal Bin Islam, and Hasan Mohammed Asiful Hoque. "Building Climate-Resilient, Migrant-Friendly Cities and Towns." Urban Climate Change Programme, International Centre for Climate Change and Development (July 2018).

Ballard, J.G. *Empire of the Sun*. London: Panther Books, 1984.

Batubara, Bosman, Michelle Kooy, and Margreet Zwarteveen. "Uneven Urbanisation: Connecting Flows of Water to Flows of Labour and Capital Through Jakarta's Flood Infrastructure." *Antipode* 50, no. 5 (April 2018): 1186–205. doi.org/10.1111/anti.12401.

Bednarik, Robert G. "The Earliest Evidence of Ocean Navigation." *International Journal of Nautical Archaeology* 26, no. 3 (August 1997): 183–91. doi.org/10.1111/j.1095-9270.1997.tb01331.x.

Begum, S., Marcel J. F. Stive, and Jim W. Hall, eds. *Flood Risk Management in Europe*. New York: Springer, 2007.

Bellwood, Peter. "Austronesian Prehistory in Southeast Asia: Homeland, Expansion and Transformation." In *The Austronesians: Historical and Comparative Perspective*, edited by Peter Bellwood, James J. Fox, and Darrell Tryon, 103–18. Canberra: ANU Press, 2006. jstor.com/stable/j.ctt2jbjx1.8.

Bicho, Nuno F., Loren G. Davis, and Jonathan A. Haws, eds. *Trekking the Shore: Changing Coastlines and the Antiquity of Coastal Settlements*. New York: Springer, 2011.

Bleakney, J. Sherman. *Sods, Soil, and Spades: The Acadians at Grand Pré and Their Dykeland Legacy*. Montreal and Kingston: McGill-Queen's University Press, 2004.

Bourasaw, Noel V. "The Calhoun Brothers of LaConner." *Skagit River Journal of History & Folklore*, July 22, 2004. skagitriverjournal.com/WestCounty/LaConner/Pioneers/Pre1890/Calhoun/Calhoun01-FirstwSullivan.html.

Büntgen, Ulf, Vladimir S. Myglan, Fredrik Charpentier Ljungqvist, Michael Cormick, Nicola Di Cosmo, Michael Sigl, Johann Jungclaus et al. "Cooling and Societal Change During the Late Antique Little Ice Age from 536 to Around 660 AD." *Nature Geoscience* 9, no. 3 (February 2016). doi.org/10.1038/ngeo2652.

Caesar, C. Julius. *Gallic War*. Boston: Loeb Classical Library, 1917.

Chatwin, Bruce. *The Songlines*. Markham, ON: Penguin Books Canada, 1987.

Chen, Delton B., Joel van der Beek, and Jonathan Cloud. "Climate Mitigation Policy as a System Solution: Addressing the Risk Cost of Carbon." *Journal of Sustainable Finance & Investment* 7, no. 3: 233–74. doi.org/10.1080/20430795.2017.1314814.

Coèdes, G. *The Indianized States of Southeast Asia*. 3rd ed. Edited by Walter F. Vella. Translated by Susan Brown Cowing. Canberra: Australian National University Press, 1975.

Crawfurd, John. *Descriptive Dictionary of the Indian Islands and Adjacent Countries*. London: Bradbury & Evans, 1856. archive.org/stream/ldpd_6769878_000/ldpd_6769878_000_djvu.txt.

Cummings, BJ. *The River That Made Seattle: A Human And Natural History of the Duwamish*. Seattle: University of Washington Press, 2020.

Dry, Sarah. *Waters of the World: The Story of the Scientists Who Unraveled the Mysteries of Our Oceans, Atmosphere, and Ice Sheets and Made the Planet Whole.* Chicago: University of Chicago Press, 2019.

Eaton, Richard M. *The Rise of Islam and the Bengal Frontier, 1204–1760.* Berkeley: University of California Press, 1993. ark.cdlib.org/ark:/13030/ft067n99v9.

El Akkad, Omar. *American War.* New York: Knopf, 2017.

Fagan, Brian. *Floods, Famines and Emperors: El Nino and the Fate of Civilizations.* New York: Basic Books, 1999.

Gaffney, Vince, Simon Fitch, and David Smith. *Europe's Lost World: The Rediscovery of Doggerland.* Research Report No. 160, Council for British Archaeology, London, November 2009. researchgate.net/publication /259639459_Europe%27s_Lost_World_The_Rediscovery_of_Doggerland.

Galili, Ehud, Jonathan Benjamin, Vered Eshed, Baruch Rosen, John McCarthy, and Liora Kolska Horwitz. "A Submerged 7000-Year-Old Village and Seawall Demonstrate Earliest Known Coastal Defence Against Sea-Level Rise." *PLOS,* December 18, 2019. doi.org/10.1371/journal.pone.0222560.

Gates, Bill. *How to Avoid a Climate Disaster: The Solutions We Have and the Breakthroughs We Need.* New York: Knopf, 2021.

Gernet, Jacques. *A History of Chinese Civilization.* 2nd ed. Translated by J.R. Foster and Charles Hartman. Cambridge: Cambridge University Press, 1996.

Ghosh, Amitav. *The Great Derangement: Climate Change and the Unthinkable.* Chicago: University of Chicago Press, 2017.

———. *Gun Island.* Toronto: Penguin Random House Canada, 2019.

———. *The Hungry Tide.* Toronto: Penguin Canada, 2005.

Goldstein, Joshua S., and Staffan A. Qvist. *A Bright Future: How Some Countries Have Solved Climate Change and the Rest Can Follow.* New York: Public Affairs, 2019.

Groeskamp, Sjoerd, and Joakim Kjellsson. "NEED: The Northern European Enclosure Dam for If Climate Change Mitigation Fails." *Bulletin of the American Meteorological Society* 101, no. 7 (2020): E1174–89. journals.ametsoc .org/view/journals/bams/101/7/bamsD190145.xml.

Harvey, M.J. "Salt Marshes of the Maritimes." *Nature Canada* 2, no. 2 (1973): 22–26.

Hasan, Shahnoor, Jaap Evers, and Margreet Zwarteveen. "The Transfer of Dutch Delta Planning Expertise to Bangladesh: A Process of Policy Translation." *Environmental Science and Policy* (December 2019). doi.org/10.1016/j.envsci .2019.11.001.

Hatvany, Matthew. *Marshlands: Four Centuries of Environmental Change on the Shores of the St. Lawrence.* Quebec: Les Presses de la Université Laval, 2004.

Henn, Brenna M., L.L. Cavalli-Sforza, and Marcus W. Feldman. "The Great Human Expansion." *PNAS* 109, no. 44 (2012): 17758–64. doi.org/10.1073/pnas .1212380109.

Herrle, Jens O., Jörg Bollmann, Christina Gebühr, Hartmut Schulz, Rosie M. Sheward, and Annika Giesenberg. "Black Sea Outflow Response to Holocene Meltwater Events." *Scientific Reports* 8 (2018): nature.com/articles/s41598-018-22453-z.

Hershkovitz, Israel, Gerhard W. Weber, Rolf Quam, Mathieu Duval, Rainer Grün, Leslie Kinsley, Avner Ayalon et al. "The Earliest Modern Humans Outside Africa." *Science* 359, no. 6374 (January 2018): 456–59. science.org/doi/10.1126/science.aap8369.

Huang, Jon, Claire O'Neill, and Hiroko Tabuchi. "Bitcoin Uses More Electricity Than Many Countries. How Is That Possible?" *New York Times*, September 3, 2021. nytimes.com/interactive/2021/09/03/climate/bitcoin-carbon-footprint-electricity.html.

Jarvis, Stephen, Olivier Deschenes, and Akshaya Jha. "The Private and External Costs of Germany's Nuclear Phase-Out." Working Paper 26598. The National Bureau of Economic Research, December 2019. nber.org/papers/w26598.

Jie, Yin, Zhan-e Yin, Xiao-meng Hu, Shi-yuan Xu, Jun Wang, Zhi-hua Li, Hai-dong Zhong, and Fu-bin Gan. "Multiple Scenario Analyses Forecasting the Confounding Impacts of Sea Level Rise and Tides from Storm Induced Coastal Flooding in the City of Shanghai, China." *Environmental Earth Sciences* 63 (2011): 407–14. doi.org/10.1007/s12665-010-0787-9.

J.S. Marshall Radar Observatory. *The Stormy Weather Group*. McGill University. Archived from the original July 6, 2011. web.archive.org/web/20110706185139/radar.mcgill.ca/who-we-are/history.html.

Kehoe, Marsely L. "Dutch Batavia: Exposing the Hierarchy of the Dutch Colonial City." *Journal of Historians of Netherlandish Art* 7, no. 1 (Winter 2015). doi.org/10.5092/jhna.2015.7.1.3.

Keith, David. "What's the Least Bad Way to Cool the Planet?" *New York Times*, October 1, 2021. nytimes.com/2021/10/01/opinion/climate-change-geoengineering.html.

Khanna, Parag. *Move: The Forces Uprooting Us*. Toronto: Scribner, 2021.

Kharecha, Pushker A., and James E. Hansen. "Prevented Mortality and Greenhouse Gas Emissions from Historical and Projected Nuclear Power." *Environmental Science and Technology* 47, no. 9 (March 2013): 4889–95. dx.doi.org/10.1021/es3051197.

Kimmelmann, Michael. "The Dutch Have Solutions to Rising Seas. The World Is Watching." *New York Times*, June 15, 2017. nytimes.com/interactive/2017/06/15/world/europe/climate-change-rotterdam.html.

Leary, Jim. *The Remembered Land: Surviving Sea-Level Rise After the Last Ice Age*. London: Bloomsbury Academic, 2016.

Liu, Coco. "Shanghai Struggles to Save Itself from the Sea." *Scientific American Climate Wire*, September 27, 2011. scientificamerican.com/article/shanghai-struggles-to-save-itself-from-east-china-sea.

Lombard, Denys. *Le darrefour javanais: Essai d'histoire globale*. Paris: Édition de l'école des hautes études en sciences sociales, 2004.

Lu, Mia, and Joanna Lewis. *China and US Case Studies: Preparing for Climate Change; Shanghai: Targeting Flood Management*. Georgetown Climate Center, August 2015. georgetownclimate.org/files/report/GCC-Shanghai _Flooding-August2015.pdf.

Mach, Katharine J., and A.R. Siders. "Reframing Strategic, Managed Retreat for Transformative Climate Adaptation." *Science* 372, no. 6548 (2021): 1294–99. doi.org/10.1126/science.abh1894.

Monbiot, George. "After the Failure of COP26, There's Only One Hope for Our Survival." *Guardian*, November 14, 2021. theguardian.com /commentisfree/2021/nov/14/cop26-last-hope-survival-climate-civil -disobedience.

Mooney, Chris, Juliet Eilperin, Desmond Butle, John Muyskens, Anu Narayanswamy, and Naema Ahmed. "Countries' Climate Pledges Built on Flawed Data, *Post* Investigation Finds." *Washington Post*, November 7, 2021. washingtonpost.com/climate-environment/interactive/2021/greenhouse -gas-emissions-pledges-data/?itid=hp-top-table-main.

Norman, K., J. Inglis, C. Clarkson, J. Faith, J. Shulmeister, and D. Harris. "An Early Colonisation Pathway into Northwest Australia 70–60,000 Years Ago." *Quaternary Science Reviews* 180 (January 15, 2018): 229–39. dx.doi.org /10.1016/j.quascirev.2017.11.023.

Nunn, Patrick. *The Edge of Memory: Ancient Stories, Oral Tradition and the Post-Glacial World*. London: Bloomsbury Sigma, 2018.

Oppenheimer, Clive. *Eruptions That Shook the World*. Cambridge, England: Cambridge University Press, 2011.

Orians, Gordon H., and Judith H. Heerwagen. "Evolved Responses to Landscapes." In *The Adapted Mind: Evolutionary Psychology and the Generation of Culture*, edited by Jerome H. Barkow, Leda Cosmides, and John Tooby, 555–79. Oxford: Oxford University Press, 1992.

Pires, Tomé. *The Suma Oriental of Tomé Pires*. Translated by Armando Cortesão. London: Hakluyt Society, 1944.

Polo, Marco. *The Travels*. Translated and edited by Ronald Latham. London: Penguin Books, 1958.

Raby, Peter. *Alfred Russel Wallace: A Life*. Princeton, NJ: Princeton University Press, 2001.

Raffles, Sir Thomas Stamford. *The History of Java*. Vol. 2. 2nd ed. London: John Murray, 1830. gutenberg.org/ebooks/49843.

Rasmussen, Carol. "Glacial Rebound: The Not So Solid Earth." NASA, August 15, 2015. nasa.gov/feature/goddard/glacial-rebound-the-not-so-solid -earth.

Reynolds, Tom. "Final Report of Hanford Thyroid Disease Study Released." *Journal of the National Cancer Institute* 94, no. 14 (2002): 1046–48. doi.org /10.1093/jnci/94.14.1046.

Ritchie, Hannah. "What Are the Safest and Cleanest Sources of Energy?" Our World in Data, February 10, 2020. ourworldindata.org/safest-sources -of-energy.

Robinson, Kim Stanley. *The Ministry for the Future.* London: Orbit, 2020.

Seabrook, Jeremy. *The Song of the Shirt: The High Price of Cheap Garments, from Blackburn to Banglades.* London: Hurst, 2015.

Sebold, Kimberly R. *From Marsh to Farm: The Landscape Transformation of Coastal New Jersey.* Washington, D.C.: National Park Service, 1992.

Sengupta, Dhritiraj, Ruishan Chena, and Michael E. Meadows. "Building Beyond Land: An Overview of Coastal Land Reclamation in 16 Global Megacities." *Applied Geography* 90 (December 5, 2017). doi.org/10.1016/j.apgeog .2017.12.015.

Sharpe, Simon, and Timothy M. Lenton. "Upward-Scaling Tipping Cascades to Meet Climate Goals: Plausible Grounds for Hope." *Climate Policy* 21, no. 4 (2021): 421–33. doi.org/10.1080/14693062.2020.1870097.

Shillito, Lisa-Marie, et al. "Pre-Clovis Occupation of the Americas Identified by Human Fecal Biomarkers in Coprolites from Paisley Caves, Oregon." *Bulletin Science Advances* 6, no. 29 (July 15, 2020). advances.sciencemag.org /content/6/29/eaba6404.

Shum, Henry. "China Eco-City Tracker: Coming Clean on Shanghai's Water Worries." Asia Pacific Foundation of Canada, March 7, 2019. asiapacific.ca/blog /china-eco-city-tracker-coming-clean-shanghais-water-worries.

Simmons, Matt. "Future of Wild Salmon: Fisheries and Oceans Canada Announces Sweeping Closures to Commercial Fisheries on the Pacific Coast." *Narwhal,* July 22, 2021. thenarwhal.ca/fisheries-oceans-canada-commercial-closures.

Soderstrom, Mary. *Concrete: From Ancient Origins to a Problematic Future.* Regina: University of Regina Press, 2020.

———. *Road Through Time: The Story of Humanity on the Move.* Regina: University of Regina Press, 2017.

Spurrell, F.C.J. "Early Sites and Embankments on the Margins of the Thames Estuary." *Archaeological Journal* 42, no.1: 269–302. doi.org/10.1080/00665983 .1885.10852177.

Taylor, James. *A Sketch of the Topography and Statistics of Dacca.* Calcutta: G.H. Huttmann, Military Orphan Press, 1840. books.google.ca/books?id =6kcOAAAAQAAJ.

Taylor, John. *The Cotton Manufacture of Dacca: A Descriptive and Historical Account of the Cotton Manufacture of Dacca in Bengal.* London: John Mortimer, 1851. archive.org/stream/1851cottonmanufactureofDacca/EX.1851.212_djvu.txt.

Tuhus-Dubrow, Rebecca. "The Activists Who Embrace Nuclear Power." *New Yorker*, February 19, 2021. newyorker.com/tech/annals-of-technology/the-activists -who-embrace-nuclear-power.

Van Arsdale, P. "*Homo Erectus* — A Bigger, Smarter, Faster Hominin Lineage." *Nature Education Knowledge* 4, no. 1 (2013): 2. nature.com/scitable/knowledge/library /homo-erectus-a-bigger-smarter-97879043.

Van Emmerik, Tim. "Research: Indonesia's Ciliwung Among the World's Worst Polluted Rivers." *The Conversation*, February 20, 2020. theconversation .com/research-indonesias-ciliwung-among-the-worlds-most-polluted-rivers -131207.

Van Schoubroeck, Frank, and Harm Kool. "The Remarkable History of Polder Systems in the Netherlands." Paper presented at the International Consultation on Agricultural Heritage Systems of the 21st Century, Chennai, India, February 18, 2010. fao.org/fileadmin/templates/giahs/PDF/Dutch-Polder -System_2010.pdf.

Wallace, Alfred Russel. *The Malay Archipelago*. Vol 1, *The Land of the Orang-utan and the Bird of Paradise*. Urbana, Il: Project Gutenberg, 2013. First published in London in 1869. ebook gutenberg.org/ebooks/2530.

Watts, Jonathan. "Indonesia Announces Site of Capital City to Replace Sinking Jakarta." *Guardian*, August 26, 2019. theguardian.com/world/2019/aug/26 /indonesia-new-capital-city-borneo-forests-jakarta.

Webber, M. Bert. "How the Salish Sea Got Its Name." SeaDoc Society. Accessed July 31, 2022. seadocsociety.org/how-the-salish-sea-got-its-name.

Wei, Betty Peh-T'i. *Shanghai: Crucible of Modern China*. Hong Kong: Oxford University Press, 1990.

Weninger, Bernhard, Rick Schulting, Marcel Bradtmöller, Lee Clare, Mark Collard, Kevan Edinborough, Johanna Hilpert et al. "The Catastrophic Final Flooding of Doggerland by the Storegga Slide Tsunami." *Documenta Praehistorica* 35 (January 2008): 16. doi.org/10.4312/dp.35.1.

Yanchilina, Anastasia G., William B.F. Ryan, Jerry F. McManus, Petko Dimitrov, Dimitar Dimitrov, Krasimira Slavova, and Mariana Filipova-Marinovac. "Compilation of Geophysical, Geochronological, and Geochemical Evidence Indicates a Rapid Mediterranean-Derived Submergence of the Black Sea's Shelf and Subsequent Substantial Salinification in the Early Holocene," *Marine Geology* 383, no. 1 (January 2017): 14–34. doi.org/10.1016/j.margeo .2016.11.001.

Zhang, Cao. "Govt to Reclaim Land from Sea Despite Environment Concerns." *Global Times*, March 11, 2010. globaltimes.cn/content/511991.shtml.

IMAGE CREDITS

INDEX

ABOUT THE AUTHOR

Photo by Mann-Tremblay

Mary Soderstrom is a Montreal-based writer of fiction and non-fiction with seven works of non-fiction, three short story collections, six novels, and one children's book to her credit. *Against the Seas: Saving Civilizations from Rising Waters* is her eighteenth book and is the logical follow-up to her last, *Concrete: From Ancient Origins to a Problematic Future.*

Mary aims to reach a wide audience in her writing in order to open eyes to problems and to indicate possible solutions to them. She has traveled widely, and those experiences enrich her work, as do a lifetime of reading everything she can get her hands on and — most of all — being curious.